Family Tree of Magnetism

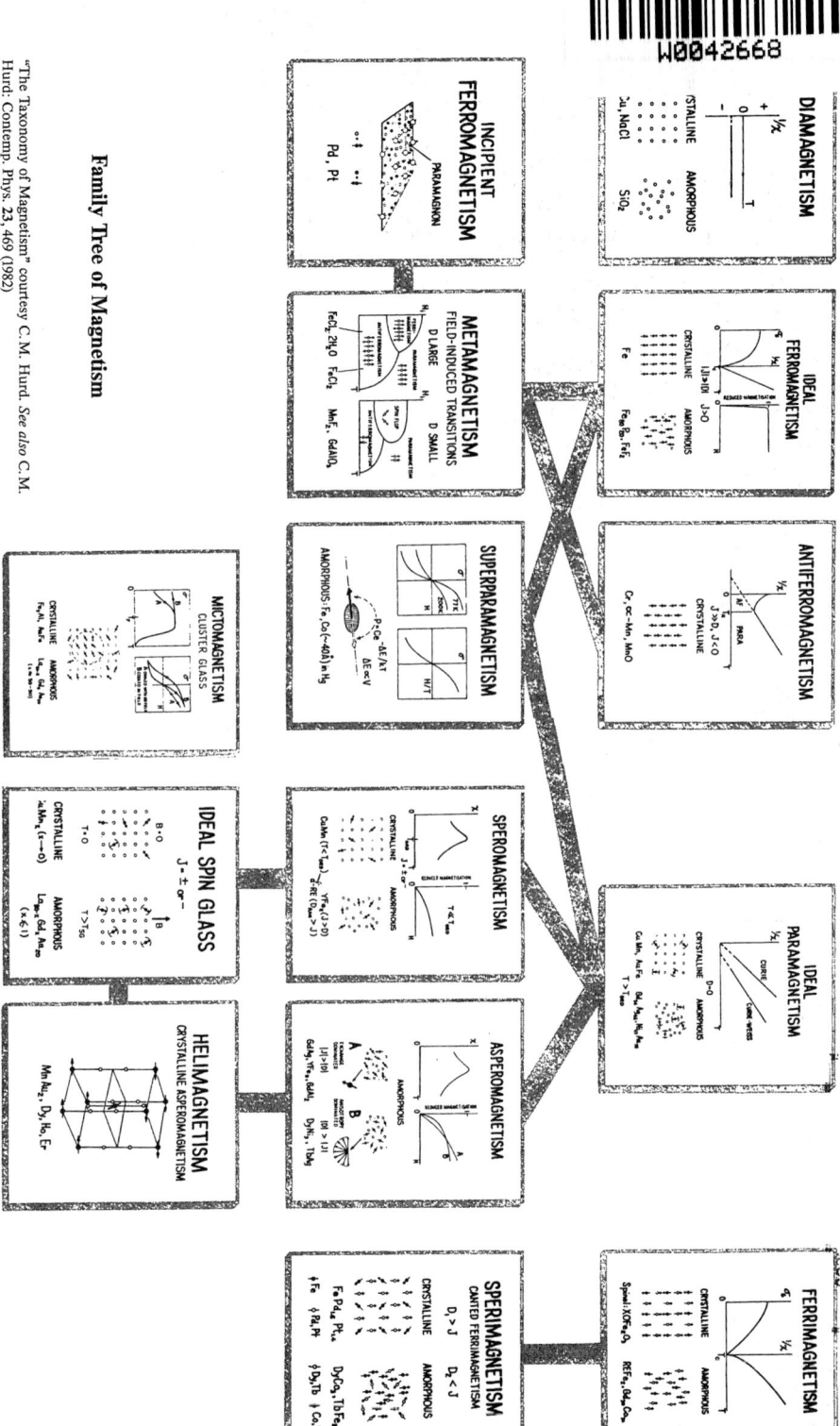

55 Springer Series in Solid-State Sciences

Edited by Peter Fulde

Springer Series in Solid-State Sciences

Editors: M. Cardona   P. Fulde   H.-J. Queisser

Volume 1–39 are listed on the back inside cover

Daniel C. Mattis

The Theory
of Magnetism II

Thermodynamics and Statistical Mechanics

With 40 Figures

Springer-Verlag
Berlin Heidelberg New York Tokyo

Professor *Daniel C. Mattis,* Ph. D.

Physics Department, University of Utah
Salt Lake City, UT 84112, USA

Series Editors:

Professor Dr. Manuel Cardona
Professor Dr. Peter Fulde
Professor Dr. Hans-Joachim Queisser

Max-Planck-Institut für Festkörperforschung, Heisenbergstrasse 1
D-7000 Stuttgart 80, Fed. Rep. of Germany

ISBN-13:978-3-642-82407-4   e-ISBN-13:978-3-642-82405-0
DOI: 10.1007/978-3-642-82405-0

Library of Congress Cataloging in Publication Data. (Revised for volume II) Mattis, Daniel Charles, 1932-.
The theory of magnetism. (Springer series in solid-state sciences ; 17, 55) Includes bibliographies and
indexes. Contents: 1. Statics and dynamics – 2. [without special title] 1. Magnetism. I. Title. II. Series.
QC753.2.M37   538   81-5060

2153/3130-543210

This book is dedicated in loving memory to
Chaim Perelman

Preface

What is thermodynamics? What does statistical physics teach us? In the pages of this slim book, we confront the answers. The reader will discover that where thermodynamics provides a large scale, macroscopic theory of the effects of temperature on physical systems, statistical mechanics provides the microscopic analysis of these effects which, invariably, are the results of *thermal disorder*.

A number of systems in nature undergo dramatic changes in aspect and in their properties when subjected to changes in ambient temperature or pressure, or when electric or magnetic fields are applied. The ancients already knew that a liquid, a solid, or a gas can represent different states of the same matter. But what is meant by "state"? It is here that the systematic study of magnetic materials has provided one of the best ways of examining this question, which is one of the principal concerns of statistical physics (alias "statistical mechanics") and of modern thermodynamics.

One century ago, Pierre Curie discovered that upon raising the temperature of a ferromagnet above a characteristic value T_c the ferromagnetism was lost, recoverable only upon subsequent cooling. It was in order to explain this phenomenon that Pierre Weiss invented the "molecular field theory" of magnetism, basing it explicitly upon van der Waals' seminal theory of non-ideal gases. But today, it is invariably Weiss' theory which is quoted and imitated, or forms the point of departure for more sophisticated theories, in such far-flung domains as many-body quantum mechanics, high-energy physics, or wherever a simple solution to a complicated system with many degrees of freedom is sought. Doubtless, the reason stems from the greater conceptual simplicity of the theory of magnetism, compared with that of gases. Magnetic theories continue to be favored today, despite the existence of a general formalism of statistical physics, which we owe to the pioneering works of A. Einstein, J.W. Gibbs, and their twentieth-century followers. But this formalism becomes totally clear only in magnetic or quasi-magnetic applications, a fact so totally admitted in the physics community that it hardly bears discussion. For example, no one need justify the study of "spin glasses" (a relatively recent concern), even if there is no practical application to such substances, because it goes without saying that they are metaphors for the structural glasses and amorphous solids, which themselves *do* have a large number of practical uses. And it also goes without saying that, however difficult the microscopic study of spin glasses may be, it is nevertheless vastly simpler than the study of the structurally random materials. And so, this book will treat a number of such "magnetic" systems not just because they are interesting or important in their own right, but also because they point the way in numerous other physical applications. 35 problems illustrate and extend the formal text.

The Theory of Magnetism I (Springer Series in Solid-State Sciences, Vol. 17), subtitled "Statics and Dynamics" treated the physics of magnetic materials and the criteria for macroscopic cooperative behavior. The present volume selects the simplest models for consideration at finite temperature, those which, like the Ising model, permit an exact analysis. The reason is the author's belief that the best way for the reader to build up physical intuition in a new subject is to observe, and participate in, the exact solution of relatively simple problems. This book provides such opportunities. At an early stage the notions of disorder (alias entropy) are introduced, followed by the important concepts (extensive vs. intensive quantities, free energy, order parameter, etc.). The molecular field theory is extended by a study of fluctuations, using the "natural" Gaussian and spherical models of T. Berlin and M. Kac.

While this text has been designed to be relatively easy to assimilate through self-study (in the event it is used in a first course in statistical physics) it is not an elementary text in the range of topics nor in the depth in which they are examined. From the simple beginnings, we go to the exact solution of a model "spin glass" which necessitates the calculation of the eigenvalues of a perfectly random matrix. Of course, we derive the spectrum of this random matrix (as simply as possible), in an attempt to keep the text self-contained. We also study the "plane rotator" model of a two-dimensional magnet, a model associated with the names of J. Kosterlitz and D. Thouless. As is often the case, a model of magnetism becomes a paradigm for a number of physical applications. In this instance: superfluidity, melting, arrays of Josephson junctions are some of them. All the elements of a rigorous solution are introduced. The Ising model, a prototype not just of interacting spins but of so many other systems in nature and in society - of all systems where discrete, individual choices influence an immediate neighborhood, and so can ultimately affect the entire thermodynamic system - is studied throughout the text, and is given an exact solution in one and two dimensions. It is also analyzed in three and higher dimensions, although the theory remains quite incomplete in those cases.

In summary, it is my intention to provide the reader with a grounding in several useful techniques and, hopefully, to present opportunities for applying them. Arguably, it is the soluble problems which determine the manner in which we perceive the insoluble ones. In the witty words of psychologist A. Maslow, "if the only tool you have is a hammer, you tend to see every problem as a nail..." So, dear reader, perhaps this new book will provide a number of tools which you will require in your kit.

Salt Lake City, February 1985   *Daniel C. Mattis*

Acknowledgments

I am grateful to C. Domb, M.E. Fisher, M.L. Glasser, E.H. Lieb, M.B. Maple, O. Symko, and a number of other colleagues who have sent me their valuable papers and allowed me to reproduce their discoveries, in the figures or in the text. I wish to thank Michael Mattis for a dedicated job of proofreading and editing of the initial drafts. There have been many changes since, and the responsibility for flaws, omissions, or confusions remains mine alone. Conversely, in the event anything significantly novel appears in this book, it requires - and deserves - acknowledgment to the National Science Foundation for support under research grant DMR-81-06223.

Contents

1. Introduction and Guide to This Text

Although it is generally the case that thermodynamic concepts can be derived from detailed statistical mechanical considerations (and not vice versa) and therefore statistical mechanics is more general than thermodynamics, one can still learn a lot from thermodynamics. As one example, standard manipulations of partial derivatives [1.1] yields a relation between the specific heat in constant magnetic field, $c_B(T)$, and that at constant magnetization, $c_m(T)$, for any material:

$$c_B(T) - c_m(T) = T(\partial m/\partial T)_B^2/(\partial m/\partial B)_T \quad . \tag{1.1}$$

To derive such an expression from statistical mechanics requires: a specific model, a knowledge of the partition function, a detailed expression for the magnetization operator (the thermal average of which we denote m) and, of course, a reasonable notion of the temperature T. All this information will be made available in the first 6 sections of Chap.2, which constitute a reasonably complete derivation and discussion of Weiss' molecular-field theory (alias mean-field theory, usually abbreviated MFT) of 1907. This theory is sufficiently venerable to have had all its flaws picked over, such as its disregard of fluctuations. L.D. Landau's version which does incorporate the fluctuations is well known from the Landau-Lifschitz books, but here we shall take a different approach.

To obtain familiarity with statistical methods, it is not necessary to consider all the microscopic theories analyzed in [1.2]. Here we shall concentrate on those few models which are simple to explain, and easy to solve. Some of these pictures have, by virtue of their simplicity, a universal appeal and have re-appeared (under various guises) in field theory or in high-energy physics [1.3]. Still, it is with some reluctance that we have omitted current topics, such as the magnetism of metals, or the Kondo phenomena at finite temperature T, because of their inherent complexity or technical difficulties. (The purpose of this text is to acquaint the reader with the *methods* of statistical physics, and such applications are still the proper domain of research.) Thus, our book will be illustrated by such "elementary" models as: Ising (esp. $s = 1/2$, with occasional diversions to $s = 1,3/2,...$), XY model ($s = 1/2$), plane rotator model (alias, classical XY model), spherical model (only the average s^2 is specified). We shall see that however simple the rules of the game, the winning strategies involve such diverse elements as nonelementary integrals (elliptic, Bessel functions, generalized Watson integrals), Fermion field-theoretic representations of Pauli's operators, etc.

In view of the large amount of information which might appear unusual to a neophyte, and which is contained in the following chapters, we shall now examine their contents in somewhat more detailed fashion than usual, virtually section-by-section.

Section 2.1. The laws of large numbers are exploited in counting configurations (the statistics of statistical mechanics), notions of temperature and free energy are introduced and a distinction drawn between extensive and intensive quantities, albeit all ad hoc.

Section 2.2. The partition function Z is defined, and established as the basis for the calculation of thermodynamic averages. Energy, free energy, entropy, magnetization, specific heat, magnetic susceptibility, and the concept of thermodynamic limit are all defined.

Section 2.3. The concept of a molecular field is explored. One immediate consequence: a specific-heat discontinuity at a critical temperature T_c defines a change of phase, a second-order phase transition, related to disappearance of the order parameter at $T \geqslant T_c$.

Section 2.4. The notion of second-order phase transition (Ehrenfest's classification: a discontinuity in the nth derivative of the free energy defines an nth-order phase transition) is pursued. It is seen that the size of the specific heat jump is related to the magnitude s of the spins, although it is always finite in the mean-field theory (abbreviated, MFT).

Section 2.5. We find some relation between paramagnetic susceptibility and spontaneous magnetization within the MFT. A built-in tendency to hysteresis (irreversible behavior through the phase transition) is pointed out.

Section 2.6. The tendency of nearest-neighbor spins to be antiparallel, a consequence of antiferromagnetic coupling, may lead to antiferromagnetism. This phenomenon requires a conceptual generalization of MFT, as the condensed phase has lower symmetry than the physical forces, i.e., possesses distinct sublattices. The discovery of this phenomenon by Néel some 50 years ago antedated the invention of microscopic investigative tools (neutron scattering, NMR, etc.) which later confirmed it. Conceptually, it raises an important question: do there exist microscopic antiferromagnetic forces which are capable of yielding antiferromagnetism, while satisfying the tenets of MFT? *Section 2.7.* Here, we address this question; we find that long-ranged ferromagnetic-type forces cause spins to be parallel and favor MFT-type ferromagnetism. (An example is found in nature: $HoRh_4B_4$.) Long-ranged antiferromagnetic-type forces, which cause spins to be antiparallel, result in a state of total disorder, as the spins are unable to all be antiparallel to one another. This phenomenon is known as frustration. A frustrated system will exhibit the disorder characteristic of high temperatures, and does not necessarily have a condensed phase, even though its paramagnetic susceptibility may satisfy the Curie-Weiss law. Finally, we see that to explain antiferromagnets, it is necessary to assume short-ranged forces and abandon MFT.

Section 2.8. Fermi and Bose statistics are derived from first principles. The Gaussian integrals which result in the classical correspondence limit are evaluated. This section also illustrates Legendre transformations from one ensemble to another, i.e. from one form of free energy to another.

Section 2.9. In the Gaussian model, spins are treated as classical harmonic oscillators while in the spherical model, their magnitudes s^2 are constrained (on average). Both models, invented by T. Berlin and M. Kac, are soluble for arbitrary range forces. We use them to study the interplay between the dimensionality d of the space lattice on which the spins are placed, and the nature of phase transition at T_c. All thermodynamic properties are expressible in terms of the distribution of normal modes, the density-of-states (dos) function. Watson's generalized integrals enter here in a natural way, and are evaluated in arbitrary dimensions. We observe that the details of the phase transition depend not merely on spin magnitudes as in MFT, but also on d (Fig.2.11).

Section 2.10. The response of Gaussian and spherical models to an external field is investigated. The former yields the Curie-Weiss law, but misbehaves at T_c. The latter yields a significantly different law and is well-behaved at T_c. The susceptibility critical exponent γ is seen to depend on d. In finite fields, the discontinuities which characterize the phase transition all disappear, in agreement with MFT. The value of T_c in the spherical model is well below that in MFT (the result of fluctuations).

Section 2.11. The magnetic properties of the spherical-model *anti*ferromagnet are calculated, the spin-flop field (at which an external field *forces* the spins to become parallel) is obtained as a function of T.

Sections 2.12 and 2.13. The spherical model of a spin glass (all spins connected to each other by perfectly random bonds) is solved exactly, using the *dos* of a random matrix. (The *dos* of random matrices is obtained by Lanczos' method of tridiagonalization, worked out explicitly here.) One interesting consequence we find is the disappearance of the phase transition in *arbitrary* (not just homogeneous or staggered) finite fields. The calculated susceptibility is seen to be in qualitative accord with experiments on AgMn (Fig.2.16). The difficulties inherent in more realistic theories of spin glasses are alluded to.

Section 2.14. The magnetization in an *Heisenberg model* of magnetism is related to the number of magnons. The linear and nonlinear magnon theories are computed and contrasted.

Section 2.15. At one time, the *Mermin-Wagner* theorem was believed to preclude phase transitions in d = 2, but E.H. Stanley and T. Kaplan gave evidence of phase transitions in 2-component vector models of magnetism in d = 2. We now understand that phase transition in models of continuous symmetry in d = 2 are special, and illustrate this by a study of XY models in the following sections. In this section, the Mermin-Wagner theorem is proved for d = 1 and 2, for all manners of models, and is seen to only preclude *long-range order*.

Section 2.16. As a first illustration, the XY model (only the x and y components of nearest-neighbor spins interact) is solved in d = 1 by a transformation to fermion field operators. The statistical mechanics is obtained using the results of Sect.2.8. The solutions can also be used as a basis for the Hartree-Fock solution of the XYZ (Heisenberg) model, as we indicate. The *plane-rotator* model is introduced (spins *constrained* to the x-y plane) and solved by means of the *transfer-matrix* formalism. The transfer matrix is diagonalized, its eigenvalues are seen to be modified Bessel functions I_n and its eigenvalues, plane waves $\exp(in\theta)$. The largest eigenvalue is seen to yield the partition function, while the others provide a basis in which to compute correlation functions, response functions, etc.

Section 2.17. The XY and plane-rotator models are studied in d = 2 dimensions. Some of the properties of the well-known Kosterlitz-Thouless phase transition are outlined, without detailed derivation.

Section 2.18. The transfer matrix in the plane-rotator problem is studied in d = 2 dimensions. It is seen to resemble a chain of connected pendula — the sine-Gordon equation. The solution — obtained by a mapping onto the ground state of the anisotropic d = 1, s = 1/2 Heisenberg antiferromagnet — is alluded to.

4

Chapter 3 is devoted to the Ising model:

Section 3.1. The high-temperature expansion, valid in arbitrary dimensions
for arbitrary (short-ranged) forces is expressed as a cumulant expansion,
expressible in terms of graphs. In the Ising model, both high-temperature
and low-temperature series can be developed and, in $d = 2$ dimensions, related
to each other by duality transformations (Sect.3.2).

Section 3.3. The two-dimensional lattices and their duals are defined. T_c is
thereby determined (Table 3.1).

Section 3.4. Nearly 50 years ago, Peierls gave a proof of the persistence of
long-range order in the $d = 2$ dimensional Ising model at finite temperature.
The proof has been superseded by Onsager's exact solution of the model but,
nevertheless, the method remains useful and is described here.

Section 3.5. The nearest-neighbor interaction Ising model in $d = 1$ dimension
is solved exactly in external field, by use of the appropriate transfer-matrix
formalism. The results for ferromagnetic and antiferromagnetic couplings are
contrasted. The calculation of correlation functions in this formalism is indicated. Magnetostriction (spin-lattice coupling) is computed.

Section 3.6. The somewhat more difficult topic of the nearest-neighbor interaction Ising model in $d = 1$ dimension, in an external *transverse* field, is
studied. The spins are represented by Pauli matrices σ_n^z, while the interaction
with the external field brings in σ_n^x, which do not commute with the former.
The methods of solution of the XY model are applicable. A transformation to
fermion field operators allows the problem to be exactly soluble. An algebraic version of the *duality* transformation allow the bonds to be mapped onto
sites, and vice versa. The nontrivial consequences of next-nearest-neighbor
couplings are investigated. The critical transverse field is determined, as
is the spectrum of excited eigenvalues. The critical value is identified with
the disappearance of an energy gap, and the onset (or disappearance) of long-
range order. Finally, the time-development of certain operators and correlations is cursorily examined.

Section 3.7. A somewhat more rigorous treatment of quadratic forms of fermions is given than in the preceding section. The role of boundary conditions
is examined, as are the complications when the quadratic form is more general
than considered heretofore. A condition for the onset (or disappearance) of
the energy gap is used to calculate the critical fields when these fields, as
well as the bonds, are arbitrary (e.g., random). The result, due to Pfeuty,
is remarkable simple: the values are critical when $<\ln|B_n|> = <\log|J_n|>$, where
the B's and J's are the applied transverse fields and the nearest-neighbor
bond parameters, respectively, and $< >$ indicates averages.

Section 3.8. The one-dimensional transfer matrix of the two-dimensional Ising model is derived in several alternate forms, and discussed. It is seen to conform to the models analyzed in Sect.3.7.

Section 3.9. The solution of the transfer matrix in the preceding section is carried out in great detail. It is contrasted with that of the Ising chain in transverse field, with which it has many features in common, such as critical exponents and T_c. Superficial analogies with the Gaussian model are also examined. In Table 3.2 we give some properties of the principal 2-dimensional lattices at T_c.

Section 3.10. The calculation of the magnetic susceptibility above T_c and of the spontaneous magnetization below it require the calculation of correlation functions. The great complexity of such calculations is reduced by the use of certain theorems concerning Toeplitz matrices, which are discussed in some detail. The results indicate significant deviations from MFT in almost all the aspects of the theory!

Section 3.11. An alternative approach to the study of partition functions has been provided by Lee and Yang in their studies of the zeros of the partition function in the *complex* fugacity plane. In the magnetic problem, this leads to the introduction of complex external fields —as can only be done mathematically, of course. We re-analyze the Ising chain ($d = 1$) in light of the Lee-Yang zeros, obtaining the familiar results for this case in the form of unfamiliar expressions.

Section 3.12. Here, we have grouped miscellaneous information concerning two-dimensional Ising models. The star-triangle transformation which, augmenting duality, allows the determination of critical properties on the various lattices, is illustrated as is the most general soluble square lattice, from which all soluble nonsquare models may be obtained. Wannier's result on the antiferromagnetic triangular lattice is discussed. (This interesting example of *frustration*, in which even at $T = 0$ a significant number of bonds are not in the state of lowest energy, illuminates the theory of spin glasses. The induced disorder destroys the phase transition despite the appearance of an homogeneous, isotropic Hamiltonian.) The susceptibility of antiferromagnets is discussed in general, and is seen to differ from that in the Néel MFT theory.

Section 3.13. This section summarizes facts concerning the Ising model in 3D. Critical parameters for various three-dimensional lattices, obtained by series expansion methods, are given in Table 3.3. (The methods by which such series are analyzed include Padé approximants, ratio methods, etc., and are touched upon in this Section.) These tabulated results are contrasted with 2D (where some exact formulas are available, in addition to the series expansions). In concluding Chap.3, we briefly generalize to higher-spin Ising models which allow the study of such phenomena as crystal-field splittings which are inaccessible within the standard spin 1/2 models.

The topic of *critical phenomena* recurs within the various parts of this text, although we have chosen not to treat it separately from the other phenomena of interest. The neighborhood of a critical point is a very interesting subject of research and study, and the reader who wishes to pursue it will find several books and reviews specifically dedicated to this topic, notably *Stanley*'s monograph on scaling theory [1.4] and *Ma*'s text on the renormalization group approach [1.5]. (If reader demand justifies it, this topic may also become the subject of a chapter 4 in some future edition of the present text.) The transfer matrix approach to the evaluation of the partition functions has been stressed, because of its general usefulness and applicability. However, one notes that many problems are being analyzed today by numerical methods — series evaluations, Monte Carlo calculations, etc., which yield valuable results even in the absence of analytical or rigorous solutions [1.6]. Even when one opts for such methods, the analytical approach advocated herein should be of some use as a starting point, although the ultimate guides will necessarily be — the computing manuals!

The beginner is now advised to turn to Sect.2.1, while more advanced readers familiar with the foundations of statistical physics may wish to start at Sect.2.3 and true sophisticates, at Sect.2.9.

2. Statistical Thermodynamics

This chapter is intended as a self-contained introduction to the statistical thermodynamics of magnetic systems. To the extent possible, all concepts and quantities are derived and calculated from first principles. Within the context of simple spin systems, we define various basic quantities: the free energy F, its derivatives the internal energy U and the entropy $\mathscr{S}$, and higher derivatives such as the specific heat c and magnetic susceptibility χ. As usual, the first hurdle in generalizing the study of physical systems to finite temperature is the very definition of temperature itself; we examine this in the following section, devoted to the simplest possible model of a thermodynamic system. Sophisticated readers should skip to Sect.2.3, or beyond.

2.1 Spins in a Magnetic Field

We start our study with the magnetic analogue of an ideal gas, a single spin interacting only with a specified external magnetic field. If this study were experimental, it would be repeated a number of times and the results subjected to statistical analysis. It is advantageous for theory to emulate this procedure; therefore we consider not one, but an assembly of N similar spins, each statistically and dynamically independent of the others, and calculate the properties *on the average*, in the limit N → ∞. This is usually denoted the *thermodynamic limit*.

Let the energy of each spin pointing parallel to the field be -h and of each spin antiparallel to the field be +h, where this Zeeman splitting parameter is

$$h = \frac{1}{2} g\mu_b B \quad .\tag{2.1.1}$$

[The Landé factor g = 2 for spin and 1 for orbital angular momentum. Bohr magneton $\mu_b = e\hbar/2mc = 0.9274 \times 10^{-20}$ erg/Gauss. Thus in a magnetic field

1 Tesla = 10,000 Gauss, the energy splitting 2h is equivalent to 0.0001 eV, or (dividing by Boltzmann's constant k_B) to a temperature $T \approx 1$ K.]

If n spins are parallel and N-n antiparallel to the field, the average energy u of each spin is related to the total energy E as follows:

$$u = \frac{E}{N} = \frac{1}{N} (N - 2n)h = (1 - 2p)h \quad , \quad \text{with} \quad p \equiv n/N \quad . \tag{2.1.2}$$

To substantiate the notion of average we need specify the counting procedure. But, at first, we shall calculate the *most probable* values of p and u. This is not only the most convenient procedure, it also serves to illustrate the methodology of statistical thermodynamics.

The probability of attaining the configuration of (2.1.2) is Q(n),

$$Q(n) = P(E) \times \frac{N!}{n!(N - n)!} \times 2^{-N} \quad . \tag{2.1.3}$$

It is a *relative* probability — un-normalized as yet. P(E) is the yet-to-be-determined thermal (Maxwell-Boltzmann) factor. The binomial coefficient counts the number of distinct arrangements of N spins which can lead to precisely the energy E of (2.1.2). The last factor relates to the normalization: 2^N is the total number of distinct configurations, or partitions, of N binary variables. In this manner, we have postulated that Q(n) is the product of two distinct probabilities: P(E), the thermal a priori probability of the system having energy E and the remaining factors being the purely statistical probability of achieving the value of n required to yield E according to (2.1.2).

The factorials can be estimated with the use of Stirling's expansion:

$$\ln N! \sim N \ln N - N + \frac{1}{2} \ln (2\pi N) + \ldots \quad . \tag{2.1.4}$$

Not only is this asymptotically exact at large N, but the error is negligible even for small N (0.3% for N = 5). The statistical factor thus depends exponentially on N in leading approximation. Unless there occurs unexpected cancellation the logarithm of Q must therefore also scale with N, the size of the system.

Quantities proportional to N are denoted *extensive* to distinguish them from the *intensive* variables (temperature, pressure, field-strength, etc.) that are independent of N. Thus, ln Q will be extensive, the sum of two terms: the first, related to the energy and the second to statistics.

We calculate ln Q:

$$\ln Q = \ln P(E) - N[\ln 2 + p \ln p + (1 - p)\ln(1 - p)] - \frac{1}{2} \ln(2\pi) \tag{2.1.5}$$

9

setting $p = n/N$. The last term is smaller by a factor N than the rest, and is discarded. The remainder of the statistical contribution is extensive. This suggests the four reasonable criteria which P(E) must satisfy in order that ln Q be useful in the determination of thermodynamic properties:

1) ln P(E) must be extensive, otherwise it would not be commensurable with the statistical term in the thermodynamic limit ($N \rightarrow \infty$). [Hence, ln Q will be extensive.]

2) ln P(E) must be dimensionless, else a change of units would modify it, while leaving the statistical term invariant. [Hence, ln Q, too, will be dimensionless.]

3) P, hence ln P, must be maximal in the ground state, i.e., must favor low energies.

Requirement (1) is met by setting.ln $P = -\beta E$, where the (intensive) coefficient β must have units (energy)$^{-1}$ to satisfy criterion (2). The negative sign is chosen so that the lowest energies have higher probability than the highest ones.

The result is a rather familiar expression:

$$P(E) = \exp(-\beta E) \quad . \tag{2.1.6}$$

[Other logical possibilities (such as ln $P(E) = -\beta E^3/N^2$, with β now having dimensions (energy)$^{-3}$) will not be considered, because they ultimately fail to agree with experiment.] Finally:

4) To attain the most probable p and u, we must maximize Q.

The coefficient β is identified with temperature as follows:

$$\beta \equiv 1/kT \tag{2.1.7}$$

where Boltzmann's constant $k = 1.3806 \times 10^{-16}$ erg/K for T in degrees Kelvin. (Of course, the choice of k is what determines the scale of T. Thus $1/\beta$ is the temperature in units $k = 1$, the scale most favored by theorists.)

Thermometry depends on being able to measure the temperature of a probe, assumed in thermal equilibrium with the system under investigation. Rudimentary though it be, the discussion above permits a proof that two or more systems in thermal equilibrium share the same temperature — that is, that a measurement of T on the one establishes T for all. The proof for spin systems is suggested in Problem 2.1.

Problem 2.1. Separate the N spins into sets of n_1, n_2, ... with $\sum n_i = N$. Assign temperatures T_1, T_2, ... to these sets and a fixed total energy E. Show that Q is maximal for $T_1 = T_2 = ... = T$, i.e., show that at fixed total energy a constant temperature is most probable.

. . . .

By analogy with P(E) the logarithm of Q is an extensive, dimensionless quantity,

$$2^N Q \equiv \exp(-\beta F) = \exp(-\beta N f) \tag{2.1.8}$$

which serves to define F, the free energy, and f, the free energy per spin. We use this definition, together with (2.1.6), to eliminate P and Q from (2.1.5), obtaining:

$$f = (1 - 2p)h + kT[p \ln p + (1 - p) \ln(1 - p)] \tag{2.1.9}$$

to O(1/N). It remains to determine the correct value of p. As the fourth criterion requires f to be a minimum, df/dp = 0. Differentiation of (2.1.9) yields

$$p = \frac{1}{1 + \exp(-2\beta h)} . \tag{2.1.10}$$

With this termal equilibrium value of p we can calculate the various thermodynamic properties, such as the *internal energy* u per spin defined in (2.1.2):

$$u = -h \tanh(\beta h) , \tag{2.1.11}$$

and the free energy per spin f in (2.1.9):

$$f = -kT \ln[2 \cosh(\beta h)] \tag{2.1.12}$$

. . . .

Problem 2.2. Derive (2.1.12) by inserting (2.1.10) into (2.1.9).

. . . .

By the above arguments, we have been led to the *most probable* properties of a thermodynamic ensemble of spins, although it was the *average* properties we set out to determine. We shall next establish that these two quantities coincide in the thermodynamic limit.

Let us denote thermal averaged quantities by < >. For example, the average n is <n>. The variance $<(n - <n>)^2>$, written $<\Delta n^2>$ for typographical simplicity, is evaluated using Q(n) as a weighing factor:

$$<\Delta n^2> = \sum_n (n - <n>)^2 Q(n) / \sum_n Q(n) \tag{2.1.13}$$

as are *all* averaged quantities. To evaluate such averages, it is convenient
to expand ln Q about its optimum value at pN, with p fixed by (2.1.10):

$$Q(n) = Q(pN) \exp\left[-\frac{1}{2} A_2(n - pN)^2 + \ldots\right] \qquad (2.1.14)$$

where (...) indicates neglect of higher-order terms starting with $A_3(n - pN)^3/3!$
(they vanish in the thermodynamic limit, as the reader may wish to verify).
Thus, $Q(n)$ is a sharply peaked Gaussian, centered on the optimum value pN.
We use Stirlings's approximation to obtain A_2:

$$A_2 = |-d^2 \ln Q/dn^2|_{pN} = \frac{d^2}{dn^2}[(N - 2n)h + n \ln n + (N - n) \ln (N - n)]_{pN}$$

$$= [Np(1 - p)]^{-1} . \qquad (2.1.15)$$

Because $Q(n)$ is symmetric about pN, $<n> = pN$ by inspection. To calculate
(2.1.13) it is easiest to approximate the sums over the slowly varying sum-
mands by integrations.

$$<\Delta n^2> = \lim_{N \to \infty} \left[\frac{\int_{-Np}^{N(1-p)} dx\, x^2 \exp(-\frac{1}{2} A_2 x^2)}{\int_{-Np}^{N(1-p)} dx \exp(-\frac{1}{2} A_2 x^2)}\right] = \frac{1}{A_2} = Np(1 - p) . \qquad (2.1.16)$$

The rms fluctuation $|\Delta n|$ is $N^{\frac{1}{2}} p^{\frac{1}{2}} (1 - p)^{\frac{1}{2}}$. Thus, the fluctuations per site
$|\Delta n|/N = p^{\frac{1}{2}} (1 - p)^{\frac{1}{2}} N^{-\frac{1}{2}}$ vanish in the thermodynamic limit.

Our results imply that the intensive observable properties are virtually
free of fluctuations—that for all practical purposes, the most probable,
the optimum, and the average values all coincide. These claims are explored
further in Problem 2.3, and are the basis for an axiomatic statistical mechanics developed in the next section.

....

Problem 2.3. Prove that the fluctuations in u about its thermodynamic equilibrium (2.1.11) are only $O(N^{-\frac{1}{2}})$ and, furthermore, that the fluctuations in
f about (2.1.12) are smaller yet, $O(N^{-1})$.

....

2.2 The Partition Function

The results of the preceding section can be obtained more elegantly through
the postulational statistical mechanics of Gibbs and Einstein. We start by
separating the noninteracting particles into a number of sets, each with
energy $e_i n_i$.

The free energy is the average of terms such as,

$$f_i = e_i p_i + kT\, p_i \ln p_i + \lambda p_i \tag{2.2.1}$$

with λ a Lagrange multiplier, chosen so as to maintain

$$\sum_i p_i = 1 \quad . \tag{2.2.2}$$

Minimizing $F = \sum_i f_i$ with respect to the individual p_i:

$$p_i = \left[\exp\left(-\frac{\lambda + kT}{kT} \right) \right] \exp(-e_i/kT) \quad . \tag{2.2.3}$$

The quantity in [ ] is usually denoted $1/Z$, where Z (for *Zustandsumme*) is the *partition function*. From (2.2.2), we have

$$Z = \sum_i \exp(-\beta e_i) = \mathrm{Tr}\{\exp(-\beta H)\} \quad , \tag{2.2.4}$$

Z, replacing the Lagrange multiplier, is thus fixed. The Hamiltonian H is hereby also introduced, assuming —as is usually the case —that the energies e_i can be written as the eigenvalues of a physically appropriate Hamiltonian H. Tr, the *trace*, which is the sum over the diagonal values of an operator, is invariant and can be evaluated in any convenient representation including the one in which H is diagonal with eigenvalues e_i.

In the event that we do not know the eigenvalues of H, it is convenient to define the *density matrix* ρ:

$$\rho = Z^{-1}\, e^{-\beta H} \quad . \tag{2.2.5}$$

In this form, it satisfies the equation

$$-\partial\rho/\partial\beta = (H - \langle H\rangle_{TA})\rho \tag{2.2.6}$$

and is normalized

$$\mathrm{Tr}\{\rho\} = 1 \quad .$$

Here we have introduced the notation for thermal average $\langle\ \rangle_{TA}$. ρ can be used to compute the thermal average of any observable or operator, as follows:

$$\langle G\rangle_{TA} = \mathrm{Tr}\{G\rho\} = \mathrm{Tr}\{\rho G\} \tag{2.2.7}$$

so that, in (2.2.6), $\langle H\rangle_{TA} \equiv \mathrm{Tr}\{H\rho\}$.

We may divine that Z has a physical significance akin to Q of the preceding section, and be led to define a free energy by its logarithm. Indeed, the basis of axiomatic statistical mechanics is precisely this!

$$Z = e^{-\beta F} \equiv \mathrm{Tr}\{e^{-\beta H}\} \quad , \tag{2.2.8}$$

which serves to define the free energy F and the partition function Z in terms of a given Hamiltonian H and temperature β^{-1}.

We define the internal energy $U \equiv <H>_{TA}$. Comparing (2.2.7,8) establishes that

$$U = \frac{\partial(\beta F)}{\partial \beta} \quad , \tag{2.2.9}$$

a familiar thermodynamic identity. Suppose next, that in an applied field B, the Hamiltonian is given by $H = H_0 - BM$, where M is the *magnetization operator*. We want to know $\mathcal{M} = <M>$, the thermal average magnetization. Again (2.2.7,8) yield the answer:

$$\mathcal{M} = -\frac{\partial F}{\partial B} \quad , \tag{2.2.10}$$

another well-known thermodynamic identity.

From (2.2.9) we find

$$U = F + \beta \frac{\partial F}{\partial \beta} = F - T \frac{\partial F}{\partial T}$$

and use this to define yet another derivative of F:

$$\mathcal{S} = -\frac{\partial F}{\partial T} = (U - F)/T \quad . \tag{2.2.11}$$

$\mathcal{S}$ is defined as the *entropy*, a measure of the thermal disorder. This aspect is explored further in Problem 2.4.

. . . .

Problem 2.4. Using the definition of ρ in (2.2.5) and (1.2.4-11), show that:

$$\mathcal{S} = -k \, \text{Tr}\{\rho \, \ln\rho\} \quad .$$

. . . .

Problem 2.4 is related to the third law of thermodynamics: unless the ground state is macroscopically degenerate, the sum rule (2.2.6) guarantees that $\mathcal{S}/N$, as calculated in Problem 2.4, vanishes at $T = 0$.

For many magnetic systems, the quantities $U, \mathcal{M}$ and $\mathcal{S}$ exhaust the thermodynamic functions related to first derivatives of F. Turning to second derivatives, the *heat capacity* C is

$$C \equiv \frac{\partial U}{\partial T} = T \frac{\partial \mathcal{S}}{\partial T} = -T \frac{\partial^2 F}{\partial T^2} \quad , \tag{2.2.12}$$

making use of (2.2.11), The heat capacity is alternately expressible as a fluctuation in the total energy:

$$C = \frac{\partial U}{\partial T} = -k\beta^2 \frac{\partial U}{\partial \beta} = -k\beta^2 \frac{\partial}{\partial \beta} \left\{ \frac{\text{Tr}[H \exp(-\beta H)]}{\text{Tr}[\exp(-\beta H)]} \right\}$$

$$= k\beta^2[<H^2> - <H>^2] = k\beta^2<(H - U)^2> \quad , \tag{2.2.13}$$

proving, incidentally, that C is non-negative.

In systems without spontaneous magnetization, when $B \to 0$ $\mathcal{M}$ will generally be proportional to the applied field B. The constant of proportionality is the *susceptibility*, χ, for which we offer a more general definition — valid for *all* systems:

$$\chi \equiv \frac{\partial \mathcal{M}(B,T,\ldots)}{\partial B} \quad . \tag{2.2.14}$$

Again, because it is a second derivative, there exists an alternative formulation as a fluctuation — this time, of the total magnetization:

$$\chi = \frac{\partial}{\partial B}\left\{\frac{Tr[M \exp(-\beta H)]}{Tr \exp(-\beta H)}\right\} = \beta<(M - \mathcal{M})^2> \quad . \tag{2.2.15}$$

The quantities: F, U, $\mathcal{S}$, $\mathcal{M}$, C, are extensive. It is often convenient to divide them by N so they no longer depend on N in the thermodynamic limit. Thus:

f = F/N	is the free energy per spin
u = U/N	is the internal energy per spin
s = $\mathcal{S}$/N	is the entropy per spin,
σ or m = $\mathcal{M}$/N	is the magnetization per spin,
c = C/N	is the *specific heat*, and
x = χ/N	is the susceptibility per spin

(used interchangeably with χ when there is no confusion possible).

A Simple Example. We now work out the properties of a spin in an external, fixed, field — the problem posed in the preceding section — using the new formalism.

The Hamiltonian of a spin one-half in an external field is a Pauli matrix, here given in a diagonal representation:

$$\mathbf{H} = -\begin{bmatrix} \frac{1}{2} & 0 \\ 0 & -\frac{1}{2} \end{bmatrix} g\mu_b B = \begin{bmatrix} -h & 0 \\ 0 & +h \end{bmatrix} \quad . \tag{2.2.16}$$

The two eigenstates are

$$\text{"up"}: \begin{bmatrix} 1 \\ 0 \end{bmatrix} \quad \text{and} \quad \text{"down"}: \begin{bmatrix} 0 \\ 1 \end{bmatrix} \quad . \tag{2.2.17}$$

To evaluate ρ, (2.2.5), we need to exponentiate the Pauli matrix. The simple identity is

$$\exp(K\sigma_z) = \mathbf{1} \cos K + \sigma_z \sinh K \tag{2.2.18}$$

15

with $K \equiv \beta h$. The matrices $\mathbf{1}$ and σ_z are

$$\mathbf{1} = \begin{bmatrix} 1 & 0 \\ 0 & 1 \end{bmatrix} \quad \text{and} \quad \sigma_z = \begin{bmatrix} 1 & 0 \\ 0 & -1 \end{bmatrix} . \tag{2.2.19}$$

The properties of the Pauli matrices are examined in many elementary quantum-mechanics texts, and in [Ref.2.1, Chap.3]. The identity (2.2.18) may be proved by summing all terms in the Taylor's series expansion, or else by verifying that both sides of the equation yield the same result on the complete set of states (2.2.17). Performing the trace over a single spin yields an ordinary function of K:

$$Z = \cosh K \, \text{Tr}\{\mathbf{1}\} + \sinh K \, \text{Tr}\{\sigma_z\} = 2 \cosh K . \tag{2.2.20}$$

For N spins, the Hamiltonian is the sum of the individual Hamiltonians; and the collective trace is, as in multidimensional integrations, the product over individual traces. Thus, for N spins,

$$Z = (2 \cosh K)^N . \tag{2.2.21}$$

The free energy and its derivative are given by

$$F = -kT \, N \, \ln(2 \cosh K) \quad \text{and} \tag{2.2.22}$$

$$U = \frac{\partial(\beta F)}{\partial \beta} = -Nh \tanh\beta h = -N(\tfrac{1}{2} g\mu_b B) \tanh(\tfrac{1}{2} \beta g\mu_b B) \tag{2.2.23}$$

in exact agreement with the results of Sect.2.1. Differentiation of F with respect to the external field B yields the magnetization:

$$\mathcal{M} = N(\tfrac{1}{2} g\mu_b) \tanh(\tfrac{1}{2} \beta g\mu_b B) \tag{2.2.24}$$

and the susceptibility

$$\chi = \frac{\partial \mathcal{M}}{\partial B} = N\beta (\tfrac{1}{2} g\mu_b)^2 \, \text{sech}^2(\tfrac{1}{2} \beta g\mu_b B) . \tag{2.2.25}$$

In the limit of zero field, the zero-field susceptibility χ_0 yields Curie's law:

$$\chi_0 = \frac{\mathcal{C}_{\frac{1}{2}}}{T} \tag{2.2.26}$$

in which $\mathcal{C}_{\frac{1}{2}}$, Curie's constant, is expressible in terms of known quantities:

$$\mathcal{C}_{\frac{1}{2}} = N(\tfrac{1}{2} g\mu_b)^2 / k . \tag{2.2.27}$$

It is amusing that the order of the limits $T \to 0$ and $B \to 0$ may *not* be inter-changed, for while χ_0 diverges at $T = 0$, the finite-field susceptibility χ (2.2.25) vanishes at $T = 0$ for any finite B, however small. We therefore note that $B = T = 0$ is a critical point in the phase space of noninteracting spins. For interacting systems, we shall find nontrivial critical points.

....

Problem 2.5. Find $\mathcal{M}$ and $\mathcal{C}$ for the following cases:

   a) spins 1 (eigenvalues $e_i = hS_i$, with $S_i = 1,0,-1$)
   b) classical dipoles (eigenvalues $e(\theta) = h\cos\theta$, with $-\pi \leqslant \theta \leqslant +\pi$)

The original calculation in case (b) dates back to *Langevin* in 1905 [2.2].

....

2.3 The Concept of the Molecular Field

Pierre Weiss invented the concept of a molecular field — or *mean* field — in 1907, by stretching an analogy with van der Waals' theory of nonideal gases. In his own words:

In 1985, in his famous Mémoire *On the Magnetic Properties of Bodies at Various Temperatures*, Pierre Curie gave the first experimental study of the magnetization of a ferromagnet, iron, as a function of field and temperature. He concluded from the curves he obtained, that 'by analogy with the hypotheses about fluids, the rapid increase of magnetization occurs when the magnetic intensity of the particles is sufficiently strong to permit them to interact'. But he also cautions against attaching too much importance to this similarity. There seems to be no reason to doubt the truth of Pierre Curie's idea, nor that many aspects of paramagnetism are to ferromagnetism what perfect gases are to dense fluids. The theory of the molecular field is an outgrowth of this idea [Ref.2.3, p.281]... modeled on van der Waals concept of internal pressure, yet different ... [Ref.2.3, p.283].

The main proposition was that the interactions — of known or unknown origins — all added to provide a single molecular field B_m, such that the *total* force on each spin was the sum of the molecular field B_m plus any external, fixed, field B applied to the sample. The equation for determining B_m is the constitutive equation of *mean field theory* (abbreviated: MFT).

   Weiss proceeded by means of a single assumption, without microscopic justification. His simple guess was that B_m would be proportional to the magnetization. For spins ½, (2.2.24) is applicable, and thus Weiss' assumption $B_m \propto \mathcal{M}$ was tantamount to:

$$B_m = B_0 \tanh[\beta b(B + B_m)] \qquad (2.3.1)$$

where B_0 is the constant of proportionality, and b is short-hand notation for $g\mu_b/2$, the ubiquitous coupling constant. Once B_m is determined by this equation, many of the results derived in the preceding sections can be taken over, the only change being the replacement of B in the various formulas by

$B + B_m$. Experiment showed that the constant of proportionality B_0 and the resultant molecular field B_m could exceed the strongest laboratory fields of some 10^5 Gauss by as much as two or three orders of magnitude! (The theory of these 'exchange forces' was treated in [2.1].) In order to obtain the results of Weiss' assumption (2.3.1) it is useful to neglect the external field B altogether, and to solve first for all the thermodynamic properties in zero applied field. Later, we can re-introduce B into our considerations, for example, in the calculation of χ_0, the susceptibility calculated in the limit $B \to 0$.

In zero external field, (2.3.1) becomes

$$B_m = B_0 \tanh(\beta b B_m) \quad . \tag{2.3.2}$$

This always possesses a trivial solution $B_m = 0$. For sufficiently large β it also admits the nontrivial solutions $\pm B_m(T)$, shown in Fig.2.1. With increasing temperature, $B_m(T)$ decreases, and ultimately vanishes when the slope of $B_0 \tanh(\beta b B_m)$ drops below 1. This occurs at temperature

$$T_c = B_0 b/k = B_0(g\mu_b/2k) \quad , \tag{2.3.3}$$

which we can identify as the Curie temperature.

How to choose between the trivial and nontrivial solutions for $T < T_c$? We can apply the criterion of maximum Z, i.e., lowest free energy F. This is one of the general principles of statistical mechanics, always invoked in case of doubt.

While the trivial solution straightaway leads to

$$Z_{triv.} = 2^N \tag{2.3.4}$$

for N spins, the nontrivial one yields

$$Z_{nontriv.} = (2 \cosh\beta b B_m)^N \quad , \tag{2.3.5}$$

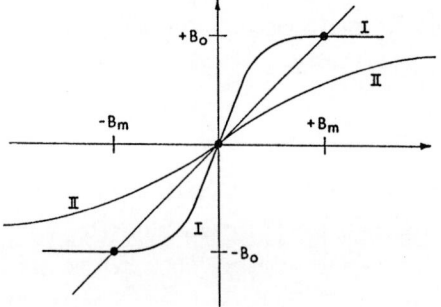

Fig.2.1. Graphical solution of (2.3.2): lhs and rhs of (2.3.2) vs internal field parameter B_m. 45° line is lhs, Curve I is rhs at $T < T_c$, displaying the nontrivial solutions at $B_m(T)$, and the solution at $B_m = 0$. Curve II is the rhs at $T > T_c$, demonstrating only the trivial solution at $B_m = 0$

according to (2.2.21). As cosh $x > 1$ for any $x \neq 0$, the nontrivial solutions are clearly preferable over the entire temperature range below T_c.

In applying (2.2.23) for the internal energy to the present analysis we must introduce a factor $\frac{1}{2}$ to eliminate double-counting the interactions. The justification, from microscopic considerations, will come later. With this single modification, (2.2.23) becomes

$$U = -N(g\mu_b/2)B_m \tanh(\beta g\mu_b B_m/2) \quad . \tag{2.3.6}$$

We eliminate the tanh factor by (2.3.2), express B_0 in terms of T_c by (2.3.3) to obtain a surprisingly simple quadratic:

$$U = -\frac{1}{2} NkT_c(B_m/B_0)^2 = -\frac{1}{2} NkT_c\sigma^2 \quad . \tag{2.3.7}$$

The physically significant quantities are: N the size of the system, T_c the experimentally determined Curie temperature, and $\sigma(T) \equiv B_m/B_0$, a function computed below.

Two other functions are of thermodynamic interest. The first is the specific heat

$$c = \frac{1}{N}\frac{\partial U}{\partial T} = -\frac{1}{2} kT_c \frac{\partial}{\partial T} \sigma^2 \tag{2.3.8}$$

and the (spontaneous) magnetization per spin, $m(T) = \mathcal{M}/N$, proportional to B_m by hypothesis, hence satisfying

$$m(T) = m(0)\sigma(T) \quad \text{with} \tag{2.3.9}$$

$$m(0) = \frac{1}{2} g\mu_b \tag{2.3.10}$$

according to (2.2.24).

Rewriting (2.3.2) in terms of $\sigma = B_m(T)/B_0$, we have:

$$\sigma = \tanh \frac{\sigma T_c}{T} \tag{2.3.11}$$

a transcendental equation for σ. While such equations are solvable by numerical iteration, or graphically (2.3.11) can also be solved by specifying a value of σ in the range $-1 \leqslant \sigma \leqslant +1$ and calculating the resulting T/T_c. Specifically,

$$\frac{T}{T_c} = \frac{2\sigma}{\ln\left(\frac{1+\sigma}{1-\sigma}\right)} = \frac{2\sigma}{2\left(\sigma + \frac{\sigma^3}{3} + \frac{\sigma^5}{5} + \ldots\right)} \quad . \tag{2.3.12}$$

The right-hand side is invariant under the change $\sigma \to -\sigma$, yielding both nontrivial roots. The result is plotted in Fig.2.2 where it is compared with

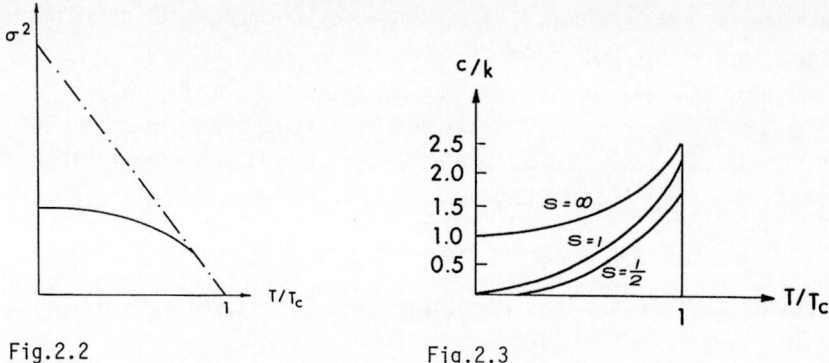

Fig.2.2

Fig.2.3

Fig.2.2. (——): solution of (2.3.12) for σ^2 vs T/T_c. (-·-·-·): linear approx-imation $\sigma^2 = 3(1-T/T_c)$, valid near T_c. σ is the order parameter

Fig.2.3. Specific heat in molecular-field theory, for $s = \frac{1}{2}$, 1, and ∞ (classi-cal limit)

the simple approximation $\sigma \sim 3^{\frac{1}{2}}(1-T/T_c)^{\frac{1}{2}}$ valid near T_c, obtained from the two leading terms in Taylor's series expansion of ln( ).

The result of the calculation for specific heat (2.3.8) is shown in Fig. 2.3, together with analogous curves for spins 1 and for classical dipoles, as previously introduced in Problem 2.5. Comparison with an experimental result on the particular substance $HoRh_4B_4$ is shown in Fig.2.9 below. The conditions for a material to satisfy the premises of mean-field theory will be examined in Sect.2.7.

2.4 Discontinuity in Specific Heat

A striking result of mean-field theory is the discontinuity in specific heat at T_c. The free energy is continuous, as it is generally required to be, and the first derivatives are all seen to vary linearly near T_c and to be con-tinuous. It is only the second derivatives —specific heat and, as we shall see, zero-field magnetic susceptibility —which are discontinuous. For this reason, the thermodynamic change of phase at T_c is denoted a *second-order phase transition*. Without any play on words being intended, this is also an *order-disorder* phase transition, from an ordered (magnetic) phase below T_c to a disordered phase above T_c. We may consider σ to be the *order parameter* which vanishes at, and above T_c.

The magnitude of the jump discontinuity can be estimated from Fig.2.3, or calculated by expanding the previous formulas for T near T_c, where $\sigma \sim 0$. It is convenient to define a dimensionless temperature t, measured from T_c:

$$t \equiv (T_c - T)/T_c \quad . \tag{2.4.1}$$

Expansion of (2.3.11 or 12) to leading order yields:

$$\sigma = (3t)^{\frac{1}{2}} \quad , \quad \text{thus} \quad U = -\frac{1}{2} NkT_c(3t) \quad \text{and} \quad c = 3k/2 \quad , \tag{2.4.2}$$

at small $t \geqslant 0$. At negative t, $\sigma = U = c = 0$. Thus, the jump discontinuity is $\Delta c = 3k/2$, independent of the model parameters T_c and B_0, but always in zero *external* field $B = 0$.

Although the jump discontinuity is generic to mean-field theories, the magnitude of the jump depends on such parameters as the magnitude of the spins (it varies from $\Delta c = 3k/2$ for spins one-half to $5k/2$ for spins $s \gg 1$) and the amount of correlations retained in the theory. Indeed, when *all* correlations are correctly taken into account in the more modern, microscopic theories which we examine later, the specific heat is seen to *diverge* near T_c as $c \sim |t|^{-\alpha}$, and it is with the calculation of such "critical exponents" as α that much of the modern research into statistical physics is concerned. Before examining these interesting new developments let us take Weiss' theory to its logical conclusions here.

We now investigate the form molecular-field theory must take when $s > \frac{1}{2}$, and the manner in which a classical limit is attained as $s \gg \frac{1}{2}$.

For a spin of given s, the eigenvalue S_z along the internal-field direction may take on any of the values: s, s-1, ..., -s+1, -s. With $b = \frac{1}{2}g\mu_b$ as before, the partition function of a single spin in a field B is an exactly summable series:

$$Z = \exp(\beta 2bsB) + \exp[\beta 2b(s-1)B] + \ldots + \exp(-\beta 2bsB)$$

$$= \frac{\sinh[\beta bB(2s+1)]}{\sinh(\beta bB)} \quad . \tag{2.4.3}$$

The magnetization (2.2.10) per spin is therefore given by

$$m(T) = \frac{\partial}{\partial B} \left\{ kT \ln \frac{\sinh[\beta bB(2s+1)]}{\sinh(\beta bB)} \right\} \quad . \tag{2.4.4}$$

The molecular-field hypothesis replaces B in the above by $B + B_m$, with a molecular field $B_m = B_0 m(T)/m(0)$, and $m(0) = 2sb$. With $B = 0$, these substitutions result in the equation

$$B_m = B_0 \mathbf{B}_s(\beta 2sbB_m) \tag{2.4.5}$$

where $\mathbf{B}_s(y)$, the function obtained by performing the differentiation indicated in (2.4.4), is the well-known *Brillouin function* [2.4]. The multiplicative constant is adjusted such that $\mathbf{B}_s(\infty) = 1$; thus the function in (2.4.5) is

$$\mathbf{B}_s(y) = \frac{1}{2s}\left[(2s+1)\coth\frac{2sy+y}{2s} - \coth\frac{y}{2s}\right] \quad . \tag{2.4.6}$$

The leading terms in expansion in a small argument y are

$$\mathbf{B}_s(y) = \frac{s+1}{3s}y - \frac{s+1}{3s}\frac{2s^2+2s+1}{30s^2}y^3 + O(y^5) \quad . \tag{2.4.7}$$

In the classical limit, $s \to \infty$, the Brillouin function reduces to the so-called Langevin function [2.2]. This was the function used by *Weiss* in his original analysis [2.3], which indeed predated the discovery of quantized spin by a decade. (We have encountered the Langevin function in the guise of $\mathcal{M}$ (Problem 2.5b). The reader should check out this limit.) Taking $\mathbf{B}_s$ to the opposite limit, $s = \frac{1}{2}$, yields the results of Sect. 2.3.

Near T_c the internal field vanishes, hence it is sufficient to equate terms linear in B_m in (2.4.5) in order to determine T_c. It is found to be

$$kT_c = \frac{2}{3}bB_0(s+1) \quad . \tag{2.4.8}$$

The expansion to $O(B_m^3)$ yields the behavior of the order parameter $\sigma = B_m/B_0 = m(T)/m(0)$ near T_c. In terms of t (which is small near T_c),

$$\sigma^2(T) = t\left(\frac{10}{3}\right)\frac{(s+1)^2}{s^2+(s+1)} - O(t^3) \quad . \tag{2.4.9}$$

For $T > T_c$, t is negative and the nontrivial roots $\sigma(T)$ are imaginary, hence inadmissible. The trivial solution, $\sigma \equiv 0$, is thus mandated above T_c.

In the expression for the internal energy $u(T)$ one can eliminate the Brillouin function and obtain a quadratic form in $\sigma(T)$, just as was done previously for $s = \frac{1}{2}$. The resulting formula for arbitrary s is

$$u = -(bsB_0)\sigma^2(T) \quad . \tag{2.4.10}$$

Just below T_c, σ^2 is given by (2.4.9) and just above it, by $\sigma \equiv 0$. Thus, there is a jump in specific heat at T_c, of magnitude:

$$\Delta c = 5k\frac{s(s+1)}{s^2+(s+1)^2} \quad . \tag{2.4.11}$$

This result, valid only in zero external field, is independent of the parameters T_c and B_0, but not of s. Δc varies from a minimum $3k/2$ for $s = \frac{1}{2}$ to a maximum $5k/2$ in the classical limit $s \to \infty$.

The molecular field theory predicts a specific heat that vanishes exponentially, as we approach $T = 0$. As the spin is increased, the region of exponential fall-off becomes less significant, until — in the classical limit of $s = \infty$ — the specific heat remains finite at $T = 0$, as explored further in Problem 2.6.

....

Problem 2.6. From the large y expansion of (2.4.6) show that in the molecular-field theory $c(T)$ approaches zero as $A \exp(-T_0/T)$ at low temperature. Express A and T_0 in terms of s and T_c. Obtain the anomalous behavior of these parameters in the classical limit (Fig.2.3), and discuss whether the point $1/s = T = 0$ is in fact a critical point of the same ilk as $B = T = 0$ for free spins.

....

The failure of $c(T)$ to vanish at absolute zero is tantamount to a failure of the third law of thermodynamics. This may be verified by integration of (2.2.12) near $T = 0$, to obtain the entropy $\mathscr{S}$ (T) in terms of the given heat capacity. It diverges, a symptom of the difficulties in classical statistical mechanics which were cured by the advent of quantum theory.

2.5 Magnetic Susceptibility and Spontaneous Magnetization

We now generalize mean-field theory to finite applied fields. Figure 2.4 shows the constitutive equation (2.3.1) plotted as function of the external field B, from which we note the salient features:

a) At fixed $T < T_c$, the curve $\sigma(B/B_0)$ is multiple-valued. The regions where σ decreases with increasing B are unphysical: Equation (2.2.15) estab-

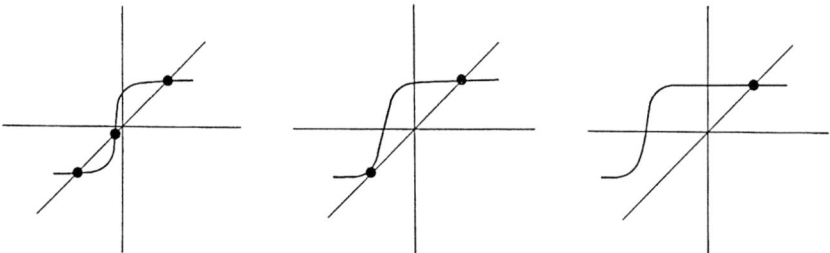

Fig.2.4. Graphical solution of (2.3.1) in nonzero external fields. *Left curve*: weak field, the 3 solutions in Fig.2.1 are slightly displaced. *Middle curve*: two roots merge at $B = B_1(T)$. *Right curve*: at fields B_1 only one root remains. See Fig.2.6 for $B_1(T)$

lished $\chi \geqslant 0$. If the applied field is increased beyond a certain value, the magnetization must jump to the upper branch. This must occur for $|B| \leqslant |B_1|$ (B_1: external field value at which $\partial\sigma/\partial B = \infty$). Thus, even the primitive mean-field theory predicts *hysteresis* —an irreversible behavior as the external field is cycled. Unfortunately, however, we cannot use this theory to study physical hysteresis, which is related to the formation and evolution of domains (regions of opposing magnetization which are created to minimize magnetostatic energy, totally ignored in the present treatment, but briefly discussed in [2.1], in connection with bubble domains [2.5,6]). In thermodynamic equilibrium, the function $\sigma(B/B_0)$ has its discontinuity strictly at $B = 0$ and is reversible, as illustrated in Fig.2.5.

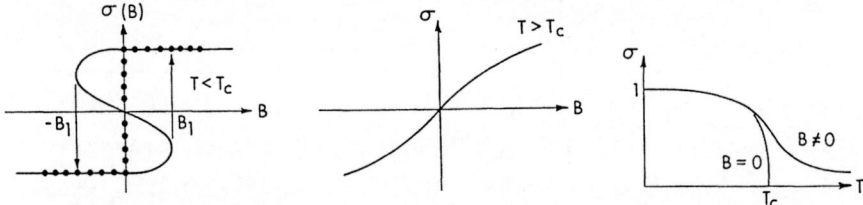

Fig.2.5. Magnetization m or order parameter σ as a function of B and T. Thermodynamic equilibrium curve is indicated by (•••), arrows show possible extent of hysteresis when B changes with time along one branch or the other

b) At fixed $B \neq 0$, $\sigma(T)$ is a smooth function, T_c being noted simply as a point of inflexion. Thermodynamic functions of magnetic systems are generally discontinuous only in zero external field. The neighborhood of $B = 0$, $T = T_c$ is denoted the critical region.

Many of these features are common to all known theories of ferromagnetism. The phase diagram we extract from mean-field theory is typical of all models that have phase transitions (not all do, as we shall see in Chap.3), and is examined in Fig.2.6. The line $\pm B_1$ indicates the extreme limit of hysteresis, although it has no significance in equilibrium thermodynamics. Figure 2.7 illustrates the zero-field susceptibility, $\chi_0 = \partial M/\partial B|_{B=0}$, as a function of T. It diverges at T_c. More precisely, mean-field theory predicts $\chi_0 \sim 1/|t|$, as we shall see shortly. It may be interesting to note here, that other models of ferromagnetism also predict a singular χ_0 at T_c. In general, one writes $\chi_0 \sim |t|^{-\gamma}$, where the critical index $\gamma \geqslant 1$. The value $\gamma = 1$ is attained only for very long-ranged forces, and it will appear that, indeed, mean-field theory is valid near T_c only for very long-ranged interactions.

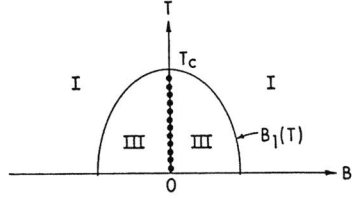

Fig.2.6. Order parameter σ is triple-valued in Regions *III*, single-valued in Region *I* although thermodynamic *equilibrium* value is unique everywhere *except* on vertical axis B = 0, T < T$_c$. Thus, 2 roots in Region *III* are always unstable or metastable

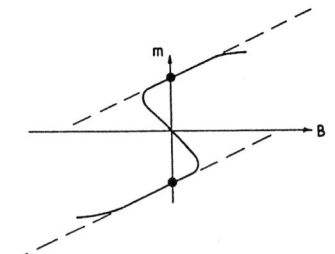 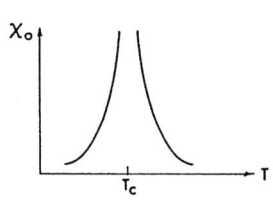 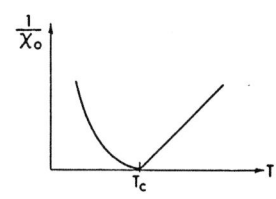

Fig.2.7. Magnetization in a ferromagnet as a function of external field. *Left graph* shows that tangents are continuous at B = 0 even though m is discontinuous. The slope ∂m/∂B = χ is the susceptibility. Plotted in *middle graph* is χ at B = 0, denoted χ$_0$, as a function of T. It is finite except at the critical temperature. For greater clarity, it is conventional to plot $1/\chi_0$ as in the *right figure*

Let us proceed with the calculations. For s = ½ at temperatures T > T$_c$ we use (2.3.1), expanding to leading order in the presumably small quantities B and B$_m$:

$$B_m = B_0(\tfrac{1}{2}\,\beta g\mu_b)(B + B_m) + O(B^3) \quad . \tag{2.5.1}$$

With (2.3.3) for T$_c$, this yields

$$B_m = -B/t \quad \text{and} \quad B + B_m = B(1 - 1/t) \tag{2.5.2}$$

The latter expression makes it easy to perform the derivative:

$$\chi_0 = \left| N(\tfrac{1}{2}\,g\mu_b)\,\frac{\partial}{\partial B}\,\tanh\!\left[\tfrac{1}{2}\,\beta g\mu_b(B + B_m)\right]\right|_{B=0}$$

$$= \frac{\mathbb{C}_{\frac{1}{2}}}{T - T_c} \quad (T > T_c) \quad , \tag{2.5.3}$$

with Curie's constant $\mathbb{C}_{\frac{1}{2}}$ previously introduced in (2.2.27).

For s > ½ we start with

$$B_m = B_0\mathbf{B}_s[\beta 2sb(B + B_m)] \quad , \tag{2.5.4}$$

and expand to first-order in the arguments, using (2.4.7). Again, this yields *precisely* (2.5.2), the s-dependence being entirely implicit in the definition

25

of $T_c = 2bB_0(s+1)/3k$, (2.4.8). Now, with $\mathcal{M} = N2bsB_m/B_0$ we find

$$\chi_0 = \frac{\mathbb{C}_s}{T - T_c} \qquad (T > T_c) \quad , \tag{2.5.5}$$

with

$$\mathbb{C}_s = Nb^2\left[\frac{4}{3} s(s+1)\right]/k = \mathbb{C}_{\frac{1}{2}}\frac{4}{3} s(s+1) \quad ,$$

the generalized Curie constant.

Below T_c one is interested not only in $\mathcal{M}_0(T)$, the spontaneous magnetization, but also in $\partial\mathcal{M}/\partial B\big|_0 \equiv \chi_0(T)$, its slope at zero field. The behavior of χ_0 is shown in Fig.2.7 and derived in Problem 2.7.

. . . .

Problem 2.7. Extend χ_0 for $s = \frac{1}{2}$ to the low-temperature region $T < T_c$ using the expansion $B_m = B_m^0 + \chi_0 B$ in (2.3.1), with $B_m^0 = $ zero-field solution of (2.3.2). Show how to obtain:

$$\chi_0 \propto \frac{(1 - \sigma^2)}{T - T_c(1 - \sigma^2)} = \begin{cases} A(T_c - T)^{-1} & T \lesssim T_c \\[2mm] \dfrac{D}{T} \exp(-T_0/T) & T \to 0 \end{cases}$$

and determine the constants A, D, T_0 in terms of T_c and $\mathbb{C}_{\frac{1}{2}}$. Next, obtain the analogous results for $s > \frac{1}{2}$.

. . . .

2.6 Antiferromagnetism

The discovery of antiferromagnetic materials tested the very foundations of molecular-field theory, for it introduced a geometrical, internal structure and a breaking of the homogeneous, isotropic symmetries of the Weiss field. Following the original explanation in terms of two interpenetrating lattices put forth by *Néel* [2.7], *Bitter* [2.8], *Van Vleck* [2.9] and others, the concepts have been extended to include spiral-, canted-, triangular, and yet-more-complex-spin arrangements revealed by neutron scattering experiments. Yet it is interesting to note how the structures were deduced prior to the developments of experimental methods which permit the direct observation of the spin correlations functions. We quote *Van Vleck*

There is ... one class of materials known as 'antiferromagnetics' in which it is quite clear that the suppression of paramagnetism is to be identified with exchange coupling. These substances ... have a susceptibility which passes through a maximum as the temperature is raised ... explained theoretically in the following way. Suppose we have a crystal whose constituent atoms can be resolved into two sublattices A and B such that the nearest neighbors of the atoms of A are atoms of B, and vice versa. The simple-cubic and body-centered cubic lattices are both of this type ... With a negative exchange integral [connecting nearest-neighbors] the exchange energy of two atoms is a minimum if their spins are antiparallel. Hence the configuration of deepest energy for the crystal as a whole is that in which the spins of sublattice A all point northward, and those of B all southward or vice versa ... Consequently a "staggered" non-vanishing molecular field can exist [2.10]

Let us pursue this idea to its logical development. For maximum simplicity, we specialize to spins $s = \frac{1}{2}$, spins of type A interacting only with B's, and vice versa.

The linear constitutive equations for two sublattices of spins one-half are written,

$$\sigma_A = \tanh\left[\frac{T_N}{T}(H - \sigma_B)\right]$$

$$\sigma_B = \tanh\left[\frac{T_N}{T}(H - \sigma_A)\right]$$

(2.6.1)

in an obvious generalization of the ferromagnetic equation (2.3.1). The applied magnetic field (in arbitrary units) is now denoted H, in order to avoid confusion with the label on one of the sublattices. σ_A and σ_B, the fraction of maximum magnetization on each sublattice, are the *order parameters*. In the absence of an external field, the order parameters on each lattice are equal in magnitude and (2.6.1) has the obvious solution $\sigma_A = -\sigma_B = \sigma$. Setting this into (2.6.1) with H = 0 reduces it to (2.3.11), an equation previously solved: see (2.3.12) and Fig.2.2 for the details.

Yet not all properties of the antiferromagnet can be related to the ferromagnet. The total magnetization, $N(\sigma_A + \sigma_B)/2$, now vanishes, so we define a staggered order parameter σ, $N(\sigma_A - \sigma_B)/2 = N\sigma$, vanishing only at or above the critical temperature. Appropriately enough, the last is denoted the *Néel temperature* T_N.

The parallel susceptibility is of great interest. It is the response to an external field along the directions of spontaneous sublattice magnetization. In a sufficiently weak external field, this quantity —denoted by $\chi_\parallel$ — is computed setting $\sigma_A = \sigma + H\chi_{\parallel A}$ and $\sigma_B = -\sigma + H\chi_{\parallel B}$, assuming σ is unaffected to $O(H^2)$. Subtracting the two equations (2.6.1), we readily verify $\chi_{\parallel A} = \chi_{\parallel B} \equiv \chi_\parallel$. Adding them yields

$$2H\chi_{\|} = \frac{\sigma + \frac{T_N}{T}H(1 - \chi_{\|})}{1 + \frac{\sigma T_N}{T}H(1 - \chi_{\|})} - \frac{\sigma - \frac{T_N}{T}H(1 - \chi_{\|})}{1 - \frac{\sigma T_N}{T}H(1 - \chi_{\|})}$$

$$= 2\frac{T_N}{T}H(1 - \chi_{\|})(1 - \sigma^2) + O(H^3) \quad . \tag{2.6.2}$$

We divide by H, then take the limit H = 0 to obtain the zero-field susceptibility

$$\chi_{0\|} = \frac{C(1 - \sigma^2)}{T + T_N(1 - \sigma^2)} \tag{2.6.3}$$

after re-expressing it in appropriate units, to agree with (2.2.25-27). σ is
the molecular-field order parameter previously calculated in (2.3.12) and
Fig.2.2.

The parallel, zero-field susceptibility given above vanishes at absolute
zero, increases to a maximum $C/2T_N$ at Néel's temperature, then decreases at
high temperature, following a modified Curie law. These properties were originally documented by *Van Vleck*

... At the absolute zero, the inhibiting effect of the powerful internal
fields on any change in alignment due to a [parallel] weak external field is
complete ...

In the same paper, he goes on to explain that an external field *perpendicular*
to the existing sublattice magnetizations would have a more important effect:

... applied perpendicular to the alternating inner fields, the external field
can still give rise to an outstanding moment, by twisting slightly the orientations of the elementary magnets [2.10].

The response to such a perpendicular field is denoted $\chi_{\perp}$ the perpendicular
susceptibility. We can calculate this quantity easily, by noting that the
perpendicular magnetization is just $\mathscr{M} \propto \sigma H_{\perp}(H_{\perp}^2 + \sigma^2)^{-\frac{1}{2}}$, hence $\chi_{0\perp}$ = const. for

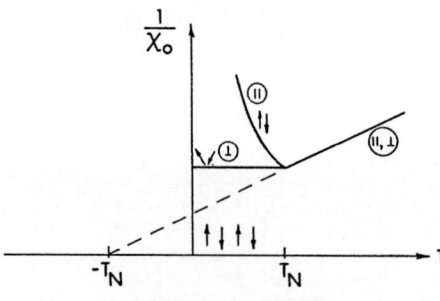

Fig.2.8. Inverse of the zero-field
longitudinal ($\|$) susceptibility and
of the zero-field transverse susceptibility ($\perp$) in an antiferromagnet subject to Néel-Weiss molecular
field theory. Both are equal and
linear above T_N. Below T_N the transverse susceptibility is constant
while the longitudinal vanishes
(hence $1/\chi_{0\|}$ diverges at T = 0).
(---) extrapolates high-temperature
line to $-T_N$

$T \leqslant T_N$. *Above* T_N the sublattice magnetizations vanish, hence χ must be isotropic and $\chi_{0\perp} = \chi_{0\parallel}$. Thus, by continuity we obtain $\chi_{0\perp} = \mathbb{C}/2T_N$ for $T \leqslant T_N$, as shown in Fig.2.8.

2.7 Short-Ranged Versus Long-Ranged Interactions

The microscopic forces in which the molecular field originates are quantum-mechanical in nature, coming in various guises depending on the specific mechanisms of electron transport in the solid. We have examined the several possibilities in [Ref.2.1, Chaps.2,4-6]. Generally, interactions are quite short-ranged, so that taking nearest- and next-nearest neighbor forces into account suffices. But there are two instances of long-ranged interactions, the first being the ubiquitous dipole-dipole electromagnetic force —weak, long-ranged, and of course, anisotropic. Its principal effect is in ferromagnetism [2.6], causing the shape of the sample to contribute to the overall energy (demagnetizing factor) and encouraging the creation of magnetic domains. It has no effect in causing a phase transition (the magnitude of the Curie temperaturature associated with purely electromagnetic forces would not exceed O(1 K)), and has no bearing on the other magnetic structures: antiferromagnetic, spiral, spin-glass, etc. with which one is often concerned. The literature on this topic is comparatively sparse [2.11]. We shall not consider the magnetostatic forces at present, but rather, a second long-ranged mechanism, the Ruderman-Kittel-Kasuya-Yosida (RKKY) mechanism, also called the indirect-exchange mechanism [2.12]. This mechanism is operative only in metals and is generally oscillatory —spins at different distances R_{ij} may be connected by either ferromagnetic or antiferromagnetic bonds, depending on the magnitude of R_{ij}. At small concentrations of electrons, this interaction becomes essentially ferromagnetic and long-ranged [2.13].

It is important to know the range and nature of the magnetic interactions because *all* microscopic theories predict quite different thermodynamic properties for short-range interactions as distinguished from long-ranged ones. This is the topic we investigate in the present section. A typical Hamiltonian includes the interactions of all pairs of spins assumed to be at points R_i of a regular lattice. Let us assume there are N such points, and that the spins $S_i = \pm 1$ are Ising $s = \frac{1}{2}$ spins. The interaction energy is thus

$$E = -\frac{1}{2} \sum_i \sum_j J_{ij} S_i S_j \quad . \tag{2.7.1}$$

Translational invariance in an homogeneous medium requires that J_{ij} depend only on the distance, $J_{ij} = J(R_{ij})$. We distinguish the range of interactions as follows:

a)   Short-ranged: $\sum_j |J(R_{ij})| < \infty$ and

b)   Long-ranged: $\sum_j |J(R_{ij})| = \infty$.

If the bonds are ferromagnetic ($J \geqslant 0$) then there is no qualitative, substantive difference between the general short-range problem (a) and a simpler model, in which only nearest-neighbor bonds are retained. If some bonds are antiferromagnetic the geometry of the lattice plays an important role. Depending on the details of the interaction and the geometry of the lattice, it may be impossible to find a ground state configuration in which all bonds J are optimized, in which case we speak of frustration. For example, the properties of a ferromagnetic *triangular* lattice are more or less independent of the range of the forces, as long as it is finite. Whereas, for antiferromagnetic coupling, it is already impossible to satisfy all the bonds just for nearest-neighbor interactions. This phenomenon, first noted by *Wannier* [2.14], is treated in the following chapter, and reappears in connection with spin glasses [2.15].

Exercise: Find the ground state of 3 spins subject to (2.7.1) in the two cases: (i) all $J_{ij} > 0$ and (ii) all $J_{ij} < 0$. Note especially, the ground state degeneracy (the number of distinguishable configurations leading to the same energy) in the latter case, if $J_{12} = J_{13} = J_{23} = J < 0$.

As a prototype of the long-ranged forces, consider a *constant* interaction $J_{ij} = J_0/N$, each spin interacting with all the others with equal strength, the factor 1/N being included so that the total energy E remains extensive. Here,

$$E = -(J_0/2N) \left(\sum_i S_i \right)^2 + J_0 N/2 . \tag{2.7.2}$$

The energy is a function of $\bar{S} = N^{-1} \sum_i S_i$ alone. We recall from Sect.2.1: let n spins be "up" and 1-n "down", such that $\bar{S} = 2p-1$, with $p = n/N$. Thus, the energy

$$E = -NJ_0\bar{S}^2/2 + J_0 N/2 = -NJ_0(2p - 1)^2/2 + J_0 N/2 \tag{2.7.3}$$

can be written as a function of p, as can the spontaneous magnetization:

$$\mathcal{M} = N(2p - 1) \tag{2.7.4}$$

and the entropy (Problem 2.4):

$$\mathscr{S} = -kN[p \ln p + (1 - p)\ln(1 - p)]$$

$$= -kN\left[\frac{1}{2}(1 + \bar{S})\ln\frac{1}{2}(1 + \bar{S}) + \frac{1}{2}(1 - \bar{S})\ln\frac{1}{2}(1 - \bar{S})\right] . \qquad (2.7.5)$$

Constructing $F(\bar{S}) = E - T\mathscr{S}$ and minimizing with respect to $\bar{S}$ (or, what is equivalent, p) we obtain a familiar result:

$$\bar{S} = \tanh\beta J_0\bar{S} \qquad (2.7.6)$$

which, for positive J_0, is easily identifiable with results of the molecular-field theory (Sect.2.3). We can define an internal field variable h_i, conjugate to the spin S_i:

$$h_i = -\frac{\partial E}{\partial S_i} = J_0\left(\bar{S} - \frac{S_i}{N}\right) . \qquad (2.7.7)$$

In the thermodynamic limit, all $h_i = J_0\bar{S}$. Thus, after we identify J_0/b as B_0, kT_c with $(J_0/b)b$, and $\bar{S}$ with σ, we recover *all* the results originally postulated by Weiss! The validity of this theory thus depends only on the range of the forces. Fortunately, there does exist an example of a metallic ferromagnet in which the range (proportional to π/k_F, the deBroglie wavelength of an electron at the Fermi surface [2.13]) may have become sufficiently long to satisfy the criteria of MFT. It is $HoRh_4B_4$, a model mean-field ferromagnet according to its discoverers [2.16]. The temperature-dependence of its spontaneous magnetization and specific heat track the theoretical curves with high precision, as shown in Fig.2.9.

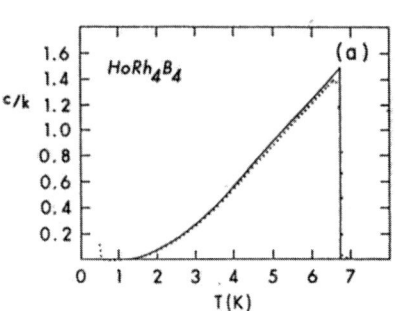 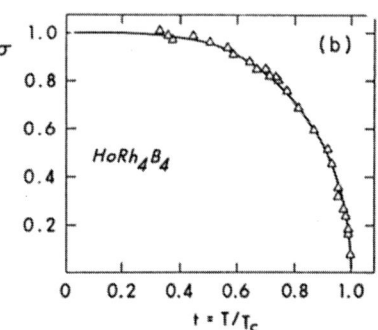

Fig.2.9a,b. An experimentally realized mean-field ferromagnetic substance showing excellent agreement between experiment and theory. (a) Temperature dependence of the specific heat [theory is (——), experimental data are the (····)]. (b) Magnetization or order parameter vs T/T_c. Taken from [2.16], where additional graphs are given for magnetostriction and electrical resistance, all in excellent accord with MFT predictions

If all ferromagnets behaved as $HoRh_4B_4$, the theory of ferromagnetism could be rather briefly disposed of. In fact, this *mean-field behavior is quite exceptional*, and we shall determine that the properties near T_c are, and should be, quite different for the more common case of short-ranged forces, and depend on dimensionality, lattice geometry, etc.

So far, we have considered $J_0 > 0$, i.e., all ferromagnetic bonds. What if J_0 were negative, would the case of all antiferromagnetic bonds lead to Néel's theory of antiferromagnetism? The resultant equation,

$$\bar{S} = -\tanh(\beta|J_0|\bar{S}) \tag{2.7.8}$$

has the unique solution $\bar{S} = 0$ at all temperatures. Looking back at (2.7.2) we see this is an example of *extreme frustration* — only half the bonds have a negative energy, the remainder have positive energy. The energy is a constant, independent of T, and one obtains $\mathscr{S}/N = k\ \ln 2$, the same entropy as for free spins. So, the answer is *no.*

Although the long-ranged antiferromagnetic interaction does not lead to any significant cooperative behavior, it does modify the magnetic susceptibility. Extending (2.7.8) to include an applied external field B we find,

$$\bar{S} = \tanh \beta(B - |J_0|\bar{S}) \tag{2.7.9}$$

which now has a nontrivial solution $\bar{S}(B,T)$. Defining the Néel temperature $T_N = |J_0|/k$, we obtain for the zero-field susceptibility an expression:

$$\chi_0 = \frac{\mathfrak{C}_{\frac{1}{2}}}{T + T_N} \tag{2.7.10}$$

which has the expected form (Fig.2.8) above T_N. Unlike the results of Néel's theory, however, χ_0 here is continuous through T_N to absolute zero. Indeed, the solution of (2.7.9) is free of any singularity at T_N (or, indeed, at any other positive temperature!). We conclude that long-ranged $J_{ij} < 0$ result only in a disordered phase at all T, and not in any antiferromagnetic ordering.

Additional insight into this apparent asymmetry between positive and negative bonds is afforded by studying the eigenvalue spectrum of $J(R_{ij})$, which is given by its Fourier transform (we may think of J_{ij} as a cyclic matrix):

$$W_k \equiv \sum_n J(R_n) \exp(ik \cdot R_n) \quad . \tag{2.7.11}$$

The amplitudes $X_k = X_{-k}^*$ of the normal modes $\exp(ik \cdot R)$ are

$$X_k = N^{-\frac{1}{2}} \sum_n S_n \exp(ik \cdot R_n) \quad , \tag{2.7.12}$$

or, conversely, the original spins are themselves a sum

$$S_n = N^{-\frac{1}{2}} \sum_n X_k \exp(-ik \cdot R_n) \qquad\qquad (2.7.13)$$

over normal modes, the sum being over the N points k of the first Brillouin zone, see [Ref.2.1, p.163] for details. The total energy (2.7.1) is now brought into diagonal form,

$$E = -\frac{1}{2} \sum_k W_k |X_k|^2 \quad . \qquad\qquad (2.7.14)$$

We are still far from being able to evaluate the partition function, because the constraints $S_n = \pm 1$ (i.e., $S_n^2 = 1$) are not satisfied for arbitrary values of the normal mode amplitudes X_k. Nevertheless, it is easy to see that for long-ranged (non-integrable) ferromagnetic interactions W_k is singular at $k = 0$, thus the optimum energy comes from having X_0 as large as possible.

Conversely, for long-ranged antiferromagnetic interactions, W_k is *negative* at $k = 0$, thus $k = 0$ is singularly *disfavored* in comparison with all the other normal modes. In such a case, we expect maximum disorder and maximum entropy, precisely as we have found in the example (2.7.8).

Turning to short-ranged forces, we find the spectrum to be non-singular. It is no longer sufficient to consider one normal mode, or one order parameter. Take the example of the simple-cubic lattice, with nearest neighbor bonds J_0:

$$W_k = -2J_0(\cos k_x a + \cos k_y a + \cos k_z a) \quad . \qquad\qquad (2.7.15)$$

Even if $J_0 > 0$ and $k = 0$ yields the optimum W_k, there will be a continuum of nearby eigenvalues which will contribute at finite temperature — the spinwaves. A single-parameter mean-field theory cannot then be expected to be accurate. If $J_0 < 0$, the optimum W_k is at $k = (\pi,\pi,\pi)/a$. The resulting spin configuration is extracted from (2.7.13):

$$S_n = (-1)^n \qquad\qquad (2.7.16)$$

where $(-1)^n = +1$ on even-numbered sites and -1 on odd-numbered sites. This is precisely Néel's state, but the mean-field thermodynamics of the preceding section are not applicable, for the reason already stated: the existence of a continuum of eigenvalues in the neighborhood of the optimum value.

If we wish Néel's state to be favored, and mean-field theory to be applicable, we must contrive that (2.7.16) be uniquely, indeed singularly, favored. One such arrangement is:

$$J(R_n) = |J_0|(-1)^n/N \qquad\qquad (2.7.17)$$

in which all bonds connecting even-numbered sites to one another are ferromagnetic, all bonds connecting odd-numbered sites to one another are also
ferromagnetic, but all bonds connecting even- to odd-numbered sites are antiferromagnetic. It is intuitive that when the interactions have the structure
of the desired ground state, the resulting order parameter is strongly favored.
However, materials in which the interactions are long-ranged and oscillatory,
in precisely this manner, are perhaps even rarer than those in which the bonds
are ferromagnetic and long-ranged. To explain antiferromagnetism it will
again be necessary to invoke short-range interactions.

For the most part, the remainder of this book will be concerned with short-
ranged interactions —usually, just nearest-neighbor interactions. But there
are many kinematically different types of individual spins resulting in:
Ising models, Heisenberg models, Gaussian, spherical, Potts, and clock models,
etc.

2.8 Fermions, Bosons, and All That

Before turning to microscopic theories of cooperative phenomena, it is necessary to study the statistical mechanics of idealized particles —non-interact-
ing fermions, bosons, as well as of spins (which are neither fermions nor
bosons), and of classical systems as well. We illustrate with a couple of
examples that will prove useful later on.

2.8.1 Fermions

No more than 1 fermion may be in any one-particle quantum state. If a given
state has energy e_k, this may also be written $n_k e_k$, with $n_k = 0$ if the state
is unoccupied and $n_k = 1$ when it is occupied. The partition function Z is
therefore of the form

$$Z = Z_0 \operatorname*{tr}_{(k)} \exp(-\beta e_k n_k) = Z_0[1 + \exp(-\beta e_k)] \qquad\qquad (2.8.1)$$

with Z_0 the partition function for all states other than k, and tr{k} refers
to the sum over the two possibilities: $n_k = 0$ and 1. One may include all
other one-particle states by induction,

$$Z = \prod_k [1 + \exp(-\beta e_k)] \quad, \tag{2.8.2}$$

where the product is over *all* possible states. This formula is not applicable if the particles interact, as we have not considered the interaction energies. The thermal averaged occupation of the kth state is

$$<n_k> = \frac{0 + 1 \exp(-\beta e_k)}{1 + \exp(-\beta e_k)} = [1 + \exp(\beta e_k)]^{-1} \tag{2.8.3}$$

the familiar *Fermi function*.

2.8.2 Bosons

We need to understand Bose-Einstein (boson) statistics for spin waves. As many bosons as desired may occupy any given one-particle state. Thus, if this state has energy e_k when occupied by a single particle, it will have energy $n_k e_k$ when occupied by n_k bosons, and $n_k = 0,1,2,\ldots$. The partition function is then

$$Z = Z_0[1 + \exp(-\beta e_k) + \exp(-2\beta e_k) + \exp(-3\beta e_k) + \ldots]$$

$$= Z_0[1 - \exp(-\beta e_k)]^{-1} \tag{2.8.4}$$

for $e_k > 0$. Again, Z_0 is the partition function for the remaining states so that, by induction,

$$Z = \prod_k [1 - \exp(-\beta e_k)]^{-1} \tag{2.8.5}$$

the product being over all distinct one-particle states, each labeled by a different value of the label k. Note that if any e_k is negative, the geometric series diverges (as does Z). In such cases, alternative methods of statistical mechanics must be explored.

The analogue of the Fermi function (2.8.3) is the Bose-Einstein distribution function, obtainable from (2.8.4):

$$<n_k> = [\exp(\beta e_k) - 1]^{-1} \quad . \tag{2.8.6}$$

. . . .

Problem 2.8. Derive the above by relating $<n_k>$ to the logarithmic derivative of Z.

. . . .

2.8.3 Gaussian

There are many examples (in acoustics, classical electromagnetism, etc.) in which the energy depends on the amplitude X_k of a normal mode, assumed to be real, in the manner: $e_k X_k^2$. The partition function then takes on an aspect:

$$Z = Z_0 \int_{-\infty}^{+\infty} dX_k \exp(-\beta e_k X_k^2) = Z_0 (\pi kT/e_k)^{\frac{1}{2}} \tag{2.8.7}$$

with Z_0 refering to all normal modes other than the kth which is singled out. The average energy in the kth normal mode is thus,

$$e_k \langle X_k^2 \rangle = \frac{1}{2} kT \tag{2.8.8}$$

. . . .

Problem 2.9. Show that odd moments $\langle X_k^{2n+1} \rangle$ all vanish. Obtain the (nonvanishing) even moments by evaluating

$$\langle X_k^{2n} \rangle = Z^{-1}(-1)^n \partial^n Z / \partial (\beta e_k)^n$$

in closed form.

. . . .

Again, by induction, Z can be written explicitly in terms of all the normal modes,

$$Z = \prod_k (\pi kT/e_k)^{\frac{1}{2}} \ . \tag{2.8.9}$$

If two normal modes are degenerate, i.e., share the same e_k, we can rewrite $e_k(X_{1k}^1 + X_{2k}^2)$ compactly as $e_k |X_k|^2$, with $X_k = X_{1k} + iX_{2k}$, and $X_k^* = X_{1k} - iX_{2k}$. In the evaluation of the partition function, we must then replace integrations

$$\int dX_{1k} \int dX_{2k} \exp[-\beta e_k (X_{1k}^2 + X_{2k}^2)] \quad \text{by} \tag{2.8.10a}$$

$$\int dX_k \int dX_k^* \exp(-\beta e_k |X_k|^2) \ , \tag{2.8.10b}$$

treating X_k, X_k^* as independent variables and evaluating integrals of type (2.8.10b) explicitly, using (2.8.10a). And, just as in the case of bosons, the parameters e_k are required to be positive in order that these expressions make any sense.

2.8.4 Legendre Transformations

By modifying the Boltzmann factors, one can often cure such deficiencies as $e_k < 0$, or evaluate certain correlations more conveniently. These shifts are the equivalent of the well-known Legendre transformations in thermodynamics. As an example, let us consider what happens to the thermodynamic relations (2.2.8-15) if the energy (2.7.1) is modified in the following manner:

$$E \rightarrow E - \mu(M - \mathcal{M}) \quad . \tag{2.8.11}$$

Let us define the entity on the rhs as $\hat{E}$, and calculate

$$\hat{Z} = e^{-\beta\hat{F}} = \mathrm{Tr}\left\{e^{-\beta\hat{E}}\right\} \quad . \tag{2.8.12}$$

The parameter μ is chosen such that $<M> \equiv \mathcal{M}$, the latter being a prescribed quantity. With this device, we deal with an ensemble in which the magnetization is fixed, rather than the usual in which the magnetic field B is given. μ is not an independent parameter, for given our assumptions, it must be chosen such that $\partial\hat{F}/\partial\mu = 0$. Replacing (2.2.10) we have $\partial\hat{F}/\partial M = \mu$. [The usual free energy F is related to F by $F = \hat{F} - \mu\mathcal{M}$. In F, μ may be taken as the independent variable, hence it is seen to correspond to the usual magnetic field, and $\partial F/\partial\mu = -\mathcal{M}$ as in (2.2.10)].

Similarly, if we wished to calculate the free energy corresponding to (2.7.14), we might have to deal with some negative values of $-W_k/2$. This is not permitted in the Gaussian integrals, so one defines

$$\hat{E} = E + \mu \left(\sum_k |X_k|^2 - N\right) \quad , \tag{2.8.13}$$

choosing μ such that $(\mu - W_k/2) > 0$ for all k. The quantity in parentheses vanishes, as may be seen by squaring both sides of (2.7.13) and summing over all N sites using the identity $S_n^2 = 1$. By using the $\hat{}$ ensemble, we guarantee that this identity is satisfied *on the average*, and avoid divergent integrals as well. (This is the basis of the spherical model, which is examined in the next section.)

Such transformations are useful for fermions as well. If the total number of particles $\sum_k n_k$ is fixed, say at $\mathcal{N}$, then the modification:

$$\sum_k e_k n_k \rightarrow \sum_k e_k n_k - \mu \left(\sum_k n_k - \mathcal{N}\right) \tag{2.8.14}$$

may be advantageous. A condition $\partial\hat{F}/\partial\mu = 0$ establishes the conservation law, while for the actual calculations, it suffices to replace e_k by $e_k - \mu$. In this instance, μ is known as the "chemical potential", or the Fermi energy (and is often written ε_F).

2.9 Gaussian and Spherical Models

The preceding can be used directly to construct exact solutions of the well-known Gaussian and spherical models of *Berlin* and *Kac* [2.17]. These two related and physically well motivated models of short-ranged spin-spin interactions demonstrate many interesting features of a phase transition at T_c, the Curie temperature. They are soluble with or without an external magnetic field, and can be probed for the properties of cooperative phase transitions: the type of decay laws of correlation functions, the temperature-dependence of thermodynamic functions near T_c (i.e., the critical exponents), the nature of low-temperature excitations, etc. Unfortunately, each of these models suffers from some shortcoming which makes it unreliable in the neighborhood of T_c and vitiates the calculated results. But before we discuss the shortcomings, let us examine their positive aspects.

2.9.1 Gaussian Model

It is here assumed that the weight $P(S)dS$ for finding the nth spin S_n in the interval between S and $S+dS$ is $(2\pi)^{-\frac{1}{2}}\exp(-S^2/2)dS$, such that

$$\int_{-\infty}^{+\infty} dSP(S) = \int_{-\infty}^{+\infty} dSS^2 P(S) = 1 \quad , \quad \text{and} \quad \int_{-\infty}^{+\infty} dSS^{2m+1}P(S) = 0$$

in which respects the Gaussian model resembles the Ising model. Unlike the latter, the distribution is continuous, peaking at $S = 0$ (the Ising model is characterized by two delta-functions, one at $S = -1$ and the other at $S = 1$).

Assuming a ferromagnetic coupling J between nearest-neighbor atoms, the partition function Z and free energy F in the absence of an external field take the form

$$Z_G = \exp(-F_G/kT) = (2\pi)^{-\frac{1}{2}N} \prod_{n=1}^{N} \int dS_n \exp(-\frac{1}{2} S_n^2)$$
$$\times \left\{ \exp\left[(J/kT) \sum_{(n,m)} S_n S_m \right] \right\} \tag{2.9.1}$$

the integrations being over all N spins, the sum being over all nearest-neighbor pairs (n,m). The coordination of nearest-neighbors depends on the number of dimensions and on the type of lattice. We expect the answers to depend on both.

Writing the N spins as a vector array

$$\mathbf{S} = (S_1, S_2, \ldots, S_N)$$

we can express the total quadratic form in the exponent of (2.9.1) as

$$-\frac{1}{2} \mathbf{S} \cdot \mathbf{A} \cdot \mathbf{S}$$

with $\mathbf{A}$ a matrix array,

$$A_{n,m} = \delta_{n,m} - (J/kT)\varepsilon_{n,m} \qquad (2.9.2)$$

where $\delta_{n,m}$ is the Kronecker delta (zero unless the lattice sites R_n and R_m are equal, in which case it is 1) and $\varepsilon_{n,m}$ is 1 when R_n and R_m are nearest-neighbor sites, and zero otherwise.

With periodic boundary conditions, $A_{n,m}$ is a cyclic matrix. Cyclic matrices have plane-wave eigenvectors, so their eigenvalues are easy to obtain as the Fourier transform of a typical row or column. Explicitly, if we write the eigenvalue equation as

$$\mathbf{A} \cdot \mathbf{v} = \lambda \mathbf{v} \quad , \quad \mathbf{v} = (v_1, v_2, \ldots, v_N) \quad ,$$

we know that setting $v_n = \cos \mathbf{k} \cdot \mathbf{R}_n$ or $\sin \mathbf{k} \cdot \mathbf{R}_n$ yields the kth eigenvalue. The λ's are easily calculated, and for a d-dimensional simple-cubic lattice are simply:

$$\lambda(\mathbf{k}) = \lambda(k_1, k_2, \ldots, k_d) = 1 - (2J/kT)(\cos k_1 + \ldots + \cos k_d) \quad . \qquad (2.9.3)$$

The lattice geometry appears in this expression. In the body-centered cubic lattice (bcc), the term $(\cos k_1 + \cos k_2 + \cos k_3)$ would be replaced by $(\cos k_1 \cdot \cos k_2 \cdot \cos k_3)$. In most lattices, an expansion of these trigonometric terms about $k = 0$ yields $a_0 - a_2 k^2 + O(k^4)$, and for ferromagnets, it is only the leading powers of $\mathbf{k}$ which determine the interesting properties. As is well known, this feature permits a universal description of critical phenomena near T_c for many different models. At present, we shall remain with the rather illuminating simple-cubic lattices, but shall consider various dimensionalities.

In terms of normal modes, the prototype integral is the Gaussian already evaluated in the preceding section. The partition function thus reduces to

$$Z_G = \prod_k [\lambda(\mathbf{k})]^{-\frac{1}{2}} = \exp\left[-\frac{1}{2} \sum_k \ln\lambda(\mathbf{k})\right] = [\text{Det}|\mathbf{A}|]^{-\frac{1}{2}} \qquad (2.9.4)$$

while the free energy *per* spin is

$$F_G/N = \frac{1}{2} kT \frac{1}{N} \sum_k \ln\lambda(\mathbf{k}) \quad . \qquad (2.9.5)$$

All λ's must be positive, or the theory fails; equation (2.9.3) shows that at or below $T_G \equiv 2Jd/k$ one or more λ's change sign. Proceeding to the thermodynamic limit, we convert the expression for F_G into an ordinary integral.

In 1D it is

$$F_G/N = \frac{1}{2} kT(2\pi)^{-1} \int_{-\pi}^{+\pi} dq \ln[1 - (2J/kT) \cos q] \quad , \qquad (2.9.5a)$$

and in 2D it is

$$F_G/N = \frac{1}{2} kT(2\pi)^{-2} \int_{-\pi}^{+\pi}\!\!\int dq_1 dq_2 \ln[1 - (2J/kT)(\cos q_1 + \cos q_2)] \qquad (2.9.5b)$$

and similarly in higher dimensions:

$$F_G/N = \frac{1}{2} kT(2\pi)^{-d} \prod_{i=1}^{d} \int_{-\pi}^{+\pi} dq_i \ln\left[1 - (2J/kT) \sum_{j=1}^{d} \cos q_i\right] \quad . \qquad (2.9.5c)$$

The two-dimensional integral can be reduced to an elliptic function and the one-dimensional integral directly evaluated by an integral identity:

$$(2\pi)^{-1} \int_{-\pi}^{+\pi} dq \ln (2 \cosh x - 2 \cos q) = |x| \qquad (2.9.6)$$

often attributed to L. Onsager.

The internal energy in d dimensions is

$$u = \frac{\partial}{\partial \beta} (\beta F_G/N) = \frac{1}{2} (2\pi)^{-d} \prod_{i=1}^{d} \int_{-\pi}^{+\pi} dq_i \frac{-2J \sum_{j=1}^{d} \cos q_j}{1 - 2J\beta \sum_{j=1}^{d} \cos q_i} \quad . \qquad (2.9.7)$$

Adding and subtracting kT to the numerator, we can cast this integral into a somewhat more compact form:

$$u = \frac{1}{2} kT - \frac{1}{2} kT \, W\!\left(d, \frac{kT}{2Jd}\right) \qquad (2.9.8)$$

where W is a *generalized Watson's integral*:

$$W(d,\tau) \equiv (2\pi)^{-d} \prod_{i=1}^{d} \int_{-\pi}^{+\pi} dq_i \left(1 - \frac{1}{\tau d} \sum_{j=1}^{d} \cos q_j\right)^{-1} \quad . \qquad (2.9.9)$$

For $T \geqslant T_G$ we only need this function in the range $\tau \geqslant 1$. We first encountered this integral for the special value $\tau = 1$ in 3D in connection with bound pairs of spinwaves [Ref.2.1, Sect.5.5]. In $d = 1,2,3$ one can evaluate $W(d,\tau)$ exactly (as shown at the conclusion of this section). Certain limiting results are easy to estimate: in the limit $\tau \to 1$, $W(1,\tau) \propto (\tau - 1)^{-\frac{1}{2}}$, $W(2,\tau) \propto \ln 1/(\tau - 1)$, and for $d > 2$, $W(d,1)$ finite, decreasing with d, arriving at a final value $W(\infty,1) = 1$. For all d, $W(d,\infty) = 1$.

The Gaussian model shows a surprising agreement with the MFT. If we approximate the bond Hamiltonian

$$H_0 = -JS_0(S_1 + S_2 + \ldots + S_{2d})$$

(which describes the interactions of a given spin S_0 with its 2d nearest - neighbors) by its mean-field average,

$$H_{OMFT} = -JS_0\bar{S}2d$$

the thermal average over S_0 yields the self-consistency equation

$$\bar{S} = \tanh(\beta J 2d)\bar{S} \quad , \quad \text{with} \quad T_c = 2dJ/k \equiv T_G \quad ,$$

which has only the trivial solution for $T > T_G$, but develops a nontrivial order parameter below T_G. Of course, the Gaussian model yields a nontrivial internal energy (2.9.8) above T_G, but fails below T_G; whereas MFT yields insignificant results above T_G and nontrivial results below T_G. The two methods are thus complementary, and agree on the value of the critical temperature. Soon, we shall see that they predict the same Curie-Weiss law for the zero-field magnetic susceptibility.

2.9.2 Spherical Model

Imposing the strict condition that the partition function be evaluated on the surface of an N-sphere

$$\sum_{i=1}^{N} S_i^2 = N \tag{2.9.10}$$

can cure some of the unphysical divergence that we noted in the Gaussian model at the critical temperature, and serves to define the spherical model. This constraint will be satisfied, on the average, by the artifice of a Lagrange multiplier μ. Incorporating the terms with the Lagrange multiplier into the Hamiltonian, we define the latter as

$$H = -J \sum_{(n,m)} S_n S_m + \mu \left(\sum_n S_n^2 - N \right) \quad . \tag{2.9.11}$$

The value of μ is determined by requiring that (2.9.10) be satisfied on average or what is equivalent, $<S_n^2> = 1$ (because all the spins are equivalent by translational invariance and isotropy of the stated model).

With this particular μ (to be determined), the partition function is

$$Z_{sph} = \exp(\mu N/kT) \prod_n \int dS_n \exp[-(\mu/kT)S_n^2]\left\{\exp\left[(J/kT) \sum_{n,m} S_n S_m\right]\right\} \quad . \tag{2.9.12}$$

It is evaluated just as in the Gaussian model, the eigenvalues λ now being functions of μ as well as T and $\mathbf{k}$:

$$\lambda(\mu,\mathbf{k}) = \frac{2}{kT}\left(\mu - J \sum_{j=1}^{d} \cos k_j\right) \quad . \tag{2.9.13}$$

Hence, the free energy is

$$F_{sph}/N = -\frac{1}{2} kT \ln(2\pi) - \mu + \frac{1}{2} kT \frac{1}{N} \sum_{\mathbf{k}} \ln\lambda(\mu,\mathbf{k}) \quad , \tag{2.9.14a}$$

but is more conveniently expressed in terms of $\tau \equiv \mu/Jd$, a scaling which introduces the appropriate units into the Lagrange multiplier. The free energy is now

$$F_{sph}/N = -\frac{1}{2} kT \ln(2\pi) + \frac{1}{2} kT \ln(2Jd/kT) - Jd\tau$$

$$+ \frac{1}{2} kT \frac{1}{N} \sum_{\mathbf{k}} \ln\left(\tau - \frac{1}{d} \sum_{j=1}^{d} \cos k_j\right) \quad . \tag{2.9.14b}$$

Differentiating it with respect to τ and using the spherical condition $\partial F_{sph}/\partial\tau = 0$ we obtain

$$\frac{kT}{2\tau} \frac{1}{N} \sum_{\mathbf{k}} \frac{1}{1 - (1/\tau d) \sum_{j=1}^{d} \cos k_j} = Jd \quad . \tag{2.9.15}$$

If $\tau > 1$ the summand is nonsingular; therefore in the thermodynamic limit the sum can be replaced by an integral.

$$\frac{1}{2} kT\tau^{-1}W(d,\tau) = Jd \tag{2.9.16a}$$

or better,

$$\frac{kT}{2Jd} = \frac{\tau}{W(d,\tau)} \quad . \tag{2.9.16b}$$

42

$W(d,\tau)$ is the integral (2.9.9) introduced in connection with the Gaussian model. Whereas (2.9.16a) is a transcendental equation for $\tau(T)$ which ordinarily one solves by numerical iteration, in the form (2.9.16b) one calculates $T(\tau)$ in a straightforward manner, obtaining the curves shown (schematically) in Fig.2.10.

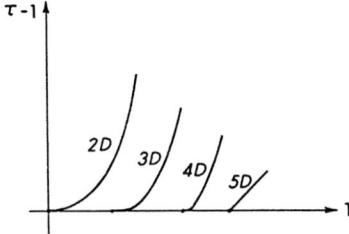

Fig.2.10. $\tau - 1$ vs T for various dimensionalities. Note that for 4D or less, the curve $\tau - 1$ intersects the horizontal axis with zero slope. At higher than 4D, the intersection is with finite slope

At the critical point $\tau = 1$, the above leads to a formula for the critical temperature,

$$kT_c(\text{sph.}) = 2Jd/W(d,1) \tag{2.9.17}$$

displayed in Table 2.1.

Table 2.1. Critical temperature in spherical model in d dimensions

d	1	2	3	4	5	6	7	...[a]	∞
$kT_c/2Jd$	0	0	0.6595	0.8068	0.871	0.900	0.916	...[a]	1

[a]For $d \gg 1$, $kT_c = 2J(d - \frac{1}{2}) + O(1/d)$

For any dimension $d > 2$, the spherical model yields a finite T_c, offering us a nontrivial low-temperature phase to investigate. We cannot merely set $\tau = 1$ in (2.9.15), as the summand now becomes singular at $k = 0$. *The integral now differs from the sum.* This deficiency is cured if we write τ in the form $1 + O(1/N)$ *before* proceeding to the thermodynamic limit. We split the sum into two contributions: the term at $k = 0$, which we treat separately, and the rest of the sum, which we now take in the thermodynamic limit, evaluating it by the corresponding integral. Thus, replacing (2.9.16a) for $T \leqslant T_c$ we have the following:

$$\frac{kT}{2N(\tau - 1)} + \frac{1}{2} kT\tau^{-1}W(d,\tau) = Jd \quad . \tag{2.9.18}$$

There is no error now in replacing τ by 1 in the second term, but the first term must be treated more carefully and requires interpretation. We define a parameter σ by

$$\frac{kT}{2Jd(\tau - 1)} \equiv N\sigma^2 \qquad (2.9.19)$$

and inserting this into (2.9.18), after eliminating $W(d,1)$ from this equation by use of (2.9.17), we obtain

$$|\sigma| = \left(1 - \frac{T}{T_c}\right)^{\frac{1}{2}} \qquad (2.9.20)$$

for the temperature-dependence of σ, which we identify as the order parameter in the spherical model. Near T_c, the critical exponent $1/2$ agrees with MFT, as computed, e.g., in Fig.2.2. The constants in the definition (2.9.19) are chosen so as to make the order parameter equal to 1 at $T = 0$.

....

Problem 2.10. Justify the identification (2.9.19) in detail, making use of (2.7.11-14) and (2.8.8) for the specific case $k = 0$. [*Hint:* see (2.9.21)ff].
....
Problem 2.11. With $F_{sph.}$ in (2.9.14a) incorporating an explicitly temperature-dependent parameter μ, prove that the thermodynamic relation $U = \langle H \rangle = \partial(\beta F)/\partial\beta$ remains valid. [*Hint:* use (2.9.15)].
....

The appearance of spontaneous ordering at $k = 0$ may be understood qualitatively as a condensation phenomenon. Let us write the thermal averaged spherical condition in the form

$$\langle X_0^2 \rangle + \sum_{k \neq 0} \langle X_k^2 \rangle = N \quad . \qquad (2.9.21)$$

Above T_c this equation is satisfied, with each $\langle X_k^2 \rangle$ being $O(1)$ (including $k = 0$), by adjusting τ. At or below T_c, τ sticks at its minimum value, $\tau = 1$, but the sum over terms $k \neq 0$ does not add up to N. To make up the deficiency in the spherical condition, one allows $\langle X_0 \rangle$ to grow according to the above:

$$\langle X_0^2 \rangle = N - \sum_{k \neq 0} \langle X_k^2 \rangle \quad ,$$

T being the only variable. Setting $\langle X_0^2 \rangle = N\sigma^2$ yields (2.9.19,20). The reader can fill in the missing details by solving Problem 2.10.

In 2D or lower dimensions, the integrals continue to approximate the sums accurately, down to $T = 0$, and there is therefore no anomalous behaviour, no T_c, and no long-range order parameter.

In calculating the thermodynamic functions such as specific heat, one must take the temperature-dependence of τ into account. Making use of Problem 2.11, we obtain

$$u = \frac{1}{2} kT - Jd\tau \qquad (2.9.22)$$

resulting in

$$c = \frac{1}{2} k \quad (T < T_c) \quad , \quad c = \frac{1}{2} k - Jd \frac{d\tau}{dT} \quad (T > T_c) \quad . \qquad (2.9.23)$$

This is related to the famous Dulong-Petit law of classical thermodynamics: $c = k/2$ per degree of freedom. For gases, this law is approximately valid at high temperatures, although it fails at low temperatures when inter-particle interactions become important. In the present context, by contrast, the Dulong-Petit law is valid only at *low* temperatures where the spherical constraint affects only the $k = 0$ mode. Above T_c it spoils this non-interaction and the correction in $d\tau/dT$ comes into play.

For $d > 4$, $d\tau/dT|_{T_c^+}$ is a finite quantity, therefore there will be a discontinuity in specific heat at T_c. For $d \leqslant 4$, $d\tau/dT$ vanishes at T_c and grows with increasing temperature, therefore there is no discontinuity in specific heat at T_c. In *all* cases, $Jd \, d\tau/dT$ approaches $k/2$ at high temperature, so that c properly vanishes at high temperature. The results are exhibited in Fig.2.11.

We have already noted in Sect.2.4 that the failure of $c(T)$ to vanish at absolute zero is tantamount to a failure of the third law of thermodynamics. In the present case, this results in an apparently flawed entropy. From its original definition, we know the entropy is a positive quantity. Using (2.9.14b) to calculate the entropy in the spherical model, we obtain

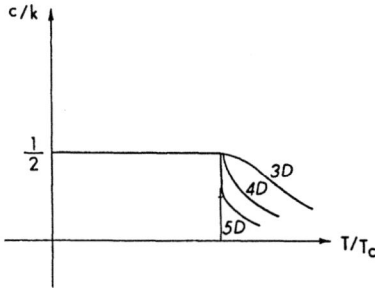

Fig.2.11. Specific heat in spherical model theory, (2.9.23). The Dulong-Petit classical law $c = k/2$ is satisfied only at $T < T_c$. At T_c the specific heat is continuous (3D), has infinite slope (4D), or is discontinuous ($d > 4$), then drops from its value at T_c^+ as $(T - T_c)^{-2}$ at very high temperatures. The limiting curve ($d = \infty$), not drawn, is qualitatively similar to the elementary MFT, Fig.2.3: finite for all $T < T_c$, zero for $T > T_c$

$$\mathscr{S}/N = -\partial f_{sph}/\partial T = \mathscr{S}_c/N - \frac{1}{2} k \ln(T_c/T) \tag{2.9.24}$$

at $T < T_c$. The change in sign in the entropy at low temperature, and its limit $-\infty$ at $T = 0$ are unphysical consequences of the model. As these defects do not occur in the Ising model, one can infer that they are related to the failure of the spins to maintain the value $S_n^2 = 1$ except on the average.

2.9.3 Watson's Integrals Generalized

The function $W(d,\tau)$ occurs in many contexts in solid-state physics. It is a special case of the lattice Green function, which in the 3D simple cubic (s.c.) lattice is given by

$$G(R,\tau) = (2\pi)^{-3} \int\!\!\int\!\!\int_{-\pi}^{+\pi} dk_1 dk_2 dk_3 \frac{\exp(i\mathbf{k} \cdot \mathbf{R})}{1 - \frac{1}{3\tau}(\cos k_1 + \cos k_2 + \cos k_3)} \tag{2.9.25}$$

in which $\mathbf{R} = (n_1, n_2, n_3)$, $n_i = 0, 1, \ldots$. The analysis of the critical phenomena in the Gaussian and spherical models depends on the value at $\mathbf{R} = 0$, which we examine here. In 1D, the integral is trivial:

$$W(1,\tau) = \frac{1}{2\pi} \int_{-\pi}^{+\pi} dk \frac{1}{1 - \frac{1}{\tau} \cos k} = \frac{\tau}{(\tau^2 - 1)^{\frac{1}{2}}} \ . \tag{2.9.26}$$

The two-dimensional integral can be evaluated as follows:

$$W(2,\tau) = \frac{1}{2\pi} \int_{-\pi}^{+\pi} dk \frac{1}{1 - \frac{1}{2\tau} \cos k} W(1, 2\tau - \cos k)$$

$$= \frac{2}{\pi} K\left(\frac{1}{\tau}\right) \tag{2.9.27}$$

where K is the complete elliptic integral of the first kind. [The properties of K are also crucial to the 2D Ising model and to the 1D Ising model in 1D; appropriate expansions and references are found at (3.6.29-31).] The three-dimensional integral $W(3,\tau)$ was first obtained by *Joyce* [2.18], who found

$$W(3,\tau) = \left(1 - \frac{3}{4} x_1\right)^{\frac{1}{2}} (1 - x_1)^{-1} (2/\pi)^2 K(k_+) K(k_-) \tag{2.9.28}$$

where

$$k_\pm^2 = \frac{1}{2} \pm \frac{1}{4} x_2 (4 - x_2)^{\frac{1}{2}} - \frac{1}{4}(2 - x_2)(1 - x_2)^{\frac{1}{2}} \quad \text{with}$$

$$x_2 = x_1/(x_1 - 1) \quad \text{and}$$

$$x_1 = \frac{1}{2} + (1/6\tau^2) - \frac{1}{2}\left(1 - \frac{1}{\tau^2}\right)^{\frac{1}{2}}[1 - (1/9\tau^2)]^{\frac{1}{2}}$$

in the region of interest, $\tau \geqslant 1$. The value at $\tau = 1$,

$$W \equiv W(3,1) = 1.516\ 386\ 059\ 151\ \ldots \tag{2.9.29}$$

recovers Watson's original result. An expansion in powers of $\tau - 1$ which is most useful near the critical point was given by Joyce as follows

$$W(3,\tau) = W - \frac{3\sqrt{3}}{2\pi}\,\varepsilon + \frac{9}{32}\left(W + \frac{6}{\pi^2 W}\right)\varepsilon^2 - \frac{3\sqrt{3}}{4\pi}\,\varepsilon^3 + \ldots \tag{2.9.30}$$

with $\varepsilon^2 \equiv 1 - 1/\tau^2$.

We use it to solve (2.9.16) near T_c

$$\tau - 1 \approx 1.68(1 - T_c/T)^2 \ , \qquad \text{for} \quad T \gtrsim T_c \ , \tag{2.9.31}$$

in 3D.

In general, one can see that $W(d,\tau)$ becomes less singular as d becomes larger, by an iteration similar to (2.9.27):

$$W(d,\tau) = \frac{1}{2\pi}\int_{-\pi}^{+\pi} dk\ \frac{1}{1 - \frac{1}{d\tau}\cos k}\ W\left(d - 1,\ \frac{\tau d - \cos k}{d - 1}\right) \tag{2.9.32}$$

with the ultimate limit $W(\infty,\tau) = 1$ in the range $1 < \tau < \infty$. In principle, one can use the power series (2.9.30) [2.18] in the above to compute $W(4,\tau)$, then use the resulting power series to compute $W(5,\tau)$, etc., but there are probably better ways to obtain these generalized Watson's integrals.

For more information on lattice Green functions, the reader is referred to *Katsura* et al. [2.19].

Finally, in the analysis of critical properties ($T \sim T_c$, $\tau \gtrsim 1$) the derivative of (2.9.16) provides a helpful equation:

$$\partial \tau / \partial T = \frac{k/2Jd}{W(d,\tau)^{-1} + W(d,\tau)^{-2}\,|\partial W(d,\tau)/\partial \tau|\tau}\ . \tag{2.9.33}$$

This is used to show that $\partial \tau/\partial T|_{T_c} = 0$ for $d \leqslant 4$.

2.10 Magnetic Susceptibility in Gaussian and Spherical Models

We factor into the partition function Z_G of the Gaussian model an interaction with an external field B:

$$Z_G(T,B) = (2\pi)^{-\frac{1}{2}N} \prod_{n=1}^{N} \int dS_n \exp\left(-\frac{1}{2} S_n^2\right)\left\{\exp\left[J/kT\right] \sum S_n S_m\right]$$

$$\times \exp\left[(b/kT)B \sum S_n\right]\right\} \quad . \tag{2.10.1}$$

A uniform shift in the N variables of integration: $S_n = S_n' + \sigma$ eliminates the term linear in the spins provided we choose

$$- \sigma + (J/kT)2d\sigma + (bB/kT) = 0 \quad . \tag{2.10.2}$$

Because $\langle S_n' \rangle = 0$, this yields the familiar Curie-Weiss law

$$\langle S_n \rangle = \sigma = B \frac{b}{k(T - T_G)} \tag{2.10.3}$$

with $T_G = 2Jd$ as before. The magnetization $m = \sigma b$ is then precisely Curie's law, (2.5.3); but the later was only valid at weak fields $B \to 0$, whereas the present result (2.10.3) is for any B, however large! This lack of saturation is a defect of the Gaussian model, a consequence of the lack of constraint on the length of the individual spins. Physically, $|\sigma|$ should not be allowed to exceed 1 however large the external field. Hopefully, the spherical model provides the cure.

For the spherical model, we similarly modify (2.9.12):

$$Z_{sph}(T,B) = \exp(\mu N/kT) \prod_{n=1}^{N} \int dS_n \exp[-(\mu/kT)S_n^2]$$

$$\times \left[\exp\left(\frac{J}{kT} \sum S_n S_m\right) \exp\left(\frac{b}{kT} \sum S_n B\right)\right] \quad . \tag{2.10.4}$$

Again, a uniform shift $S_n = S_n' + \sigma$, $dS_n = dS_n'$ eliminates terms linear in the spins, provided we choose

$$- 2\mu\sigma + J2d\sigma + bB = 0 \tag{2.10.5}$$

and therefore,

$$\langle S_n \rangle = \sigma = B \frac{b}{2\mu - 2Jd} = B \frac{b}{2Jd(\tau - 1)} \tag{2.10.6}$$

using $\mu = Jd\tau$. The value of τ, via the spherical condition, is now somewhat modified because the external field affects $\langle S_n^2 \rangle$.

$$\langle S_n^2 \rangle = \langle S_n'^2 \rangle + \sigma^2 \tag{2.10.7}$$

as the cross-terms $\langle S_n' \sigma \rangle$ vanish by symmetry. Thus, (2.9.15,16) must be modified. Simply,

$$\tau^{-1}W(d,\tau) = (1 - \sigma^2)2Jd/kT \qquad (2.10.8)$$

replaces (2.9.16). Equations (2.10.6) for $\sigma(B,\tau)$ and (2.10.8) for $\tau(B,\sigma,T)$ must be solved simultaneously —numerically of course —to yield the desired $\sigma(B,T)$ and $\tau(B,T)$.

Solving for $T(B,\tau)$ and substituting for σ by use of (2.10.6) we obtain the explicit formula:

$$kT = 2Jd\left[1 - \frac{(Bb/2Jd)^2}{(\tau - 1)^2}\right]\frac{\tau}{W(d,\tau)} \ . \qquad (2.10.9)$$

At any finite B, this yields $T(\tau)$ for all positive T, with $\tau \geqslant 1 + |Bd/2Jd|$. T is analytic in τ in these intervals, thus the phase transition at T_c is eradicated in the presence of a finite external field, as illustrated in Fig.2.12. The physical explanation for this applies to all model ferromagnets. The external field sets up long-range order even at high T, thus the order-disorder phase transition which would otherwise occur spontaneously at T_c is pushed up to T $=\infty$.

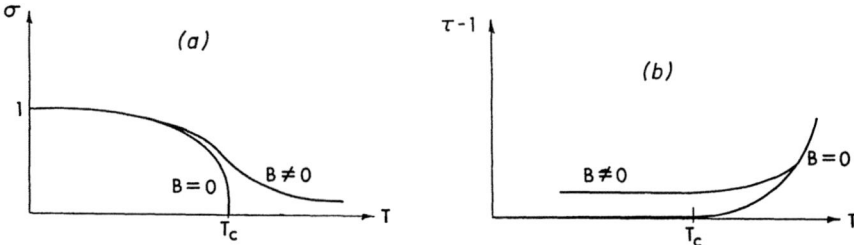

Fig.2.12a,b. Order parameters (a) σ and (b) $\tau - 1$ as functions of T in spherical model. *Lower curve* of each pair is in zero field, and shows the characteristic discontinuity at T_c. *Upper curve* shows the order parameters are smooth functions of T in finite field B $\neq 0$

Nevertheless, we can obtain the *zero-field* susceptibility by differentiating σ with respect to B in zero field, using the value of $\tau(T)$ calculated in zero field, (2.9.16). We have just obtained an estimate of this in 3D, (2.9.31), which shows $\tau - 1 \propto (T - T_c)^2$, hence by (2.10.6) $\chi_0 \propto (T - T_c)^{-2}$.

Near T_c it is customary in all theories of magnetism —not just the spherical model —to try to fit χ_0 to a power law,

$$\chi_0 = \frac{C}{(T - T_c)^\gamma} \quad , \quad T \gtrsim T_c \quad , \qquad (2.10.10)$$

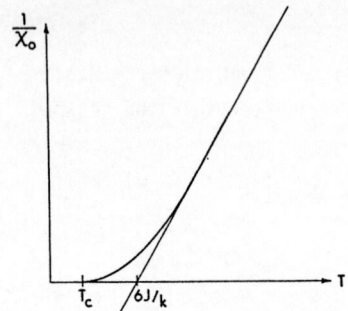

Fig.2.13. In 3D, $1/\chi_0$ for spherical model ferromagnet is parabolic near its T_c (6J/k is MFT value of T_c, shown by linear extrapolation to lie above $T_{csph} \approx 4J/k$)

where C is a constant, and γ is the susceptibility critical exponent, a quantity which depends on the symmetry of the model as well as on the number d of dimensions. For example, $\gamma = 1$ in the Gaussian model for all d, whereas $\gamma = 2$ in the spherical model for $d = 3$, and $\gamma = 1$ for the same model in $d \geqslant 5$. The susceptibility for the spherical model in 3D is plotted in Fig.2.13.

....

Problem 2.12. Obtain the correct law to replace (2.10.10) for the special case $d = 4$, in the spherical model approximation.
....

2.11 Spherical Antiferromagnet

The nearest-neighbor antiferromagnet on a bipartite lattice has precisely the same free energy as the corresponding ferromagnet in the absence of an external field. A bipartite lattice is any lattice, in any number of dimensions, which can be decomposed into two sublattices such that the sites on one sublattice (say the A sublattice) interact only with those on the other (B) sublattice. The s.c. lattice examined in the preceding sections is a special case. The transformation of the antiferromagnet proceeds explicitly, by re-defining up and down on the B-sublattice but not on the A, thus effectively changing the sign of the coupling constant J. Although this transformation is not operative in quantum Heisenberg systems (one is not allowed to invert three spin components; see [Ref.2.1, Example on p.68], it applies to Ising models, Gaussian and spherical models, and to quantum XY models. Where applicable, it implies that in the absence of external fields the thermodynamic functions are even functions of J; thus, changing the sign of J has no effect.

The important difference now comes in the magnetic properties. Changing the sign of the spin direction on alternating sites results in

$$\exp\left[\frac{b}{kT}\sum_n S_n(-1)^n B\right] \tag{2.11.1}$$

as the extra factor in the computation of the partition function in an external, homogeneous, magnetic field B. This factor differs by the alternating $(-1)^n$ from the corresponding factor in (2.10.4).

Again, a shift in the origin of the spin coordinates eliminates terms linear in the S_n in the partition function, but the shift is no longer given by (2.10.5). Rather, it is

$$S_n = S'_n + (-1)^n \sigma \quad \text{with} \tag{2.11.2}$$

$$\sigma = \frac{bB/2Jd}{\tau + 1} \quad . \tag{2.11.3}$$

Note that the net magnetization is in the direction of the applied field on *either* sublattice. The zero-field paramagnetic susceptibility is therefore,

$$\lim_{B \to 0}(\sigma/B) = \chi_0 = \frac{D}{\tau + 1} \quad (D, \text{ constant}; \; T \geqslant T_N) \tag{2.11.4}$$

with τ the solution of (2.9.16), the same as for a ferromagnet. Because of the +1 in the denominator, the susceptibility does not now diverge at T_c (which, properly, must be relabeled T_N in the present case). Morever, it is constant below T_N, having a value

$$\chi_0(T) = \chi_0(T_N) = \frac{D}{2} \quad , \quad T \leqslant T_N \quad . \tag{2.11.5}$$

Proceeding to finite external fields, we compute the magnitude of the spin-flop field, at which the external field is sufficiently large to force the spins into parallelism despite the tendency to antiparallelism due to the antiferromagnetic coupling. Equation (2.10.8) is un-modified, but with $(\tau + 1)$ replacing $(\tau - 1)$ in the evaluation of σ, we obtain a new equation of state replacing (2.10.9):

$$kT = 2Jd\left(1 - \frac{(Bb/2Jd)^2}{(\tau + 1)^2}\right)\frac{\tau}{W(d,\tau)} \quad . \tag{2.11.6}$$

If $|B|$ exceeds $B_0 \equiv 4Jd/b$ there will be only a single phase, the paramagnetic. For any $|B|$ less than this, the critical temperature is finite and is calculated to be

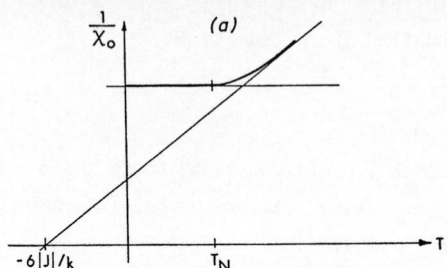 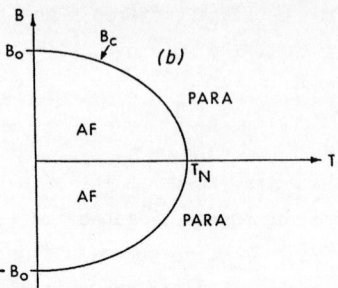

Fig.2.14. (a) Inverse susceptibility in spherical model antiferromagnet, showing constant portion for $T < T_N$, the linearly extrapolated MFT estimate of T_N, and the asymptotic Curie-Weiss-Néel behavior at high T. (b) B_C, the spin-flop field, as a function of T. This curve marks a thermodynamic phase boundary; all thermodynamic functions and order parameters are discontinuous or have discontinuous derivatives on crossing this curve. Inside the curve, the behavior is that of an AF; outside, it is paramagnetic

$$kT_N(B) = 2Jd\left[1 - \frac{1}{4}(Bb/2Jd)^2\right]/W(d,1)$$

$$= kT_{\bar{N}}(0)[1 - (B/B_0)^2] \qquad (2.11.7)$$

by setting $\tau = 1$ in the preceding. An antiferromagnetic (AF) phase thus persists at any given $T < T_N(B)$, or conversely, for $|B| < B_c(T)$, where $B_c(T)$ the spin-flop field can be obtained from (2.11.7). The phase diagram and magnetic properties are sketched in Fig.2.14.

Physically, it is simple to see why the behavior differs from the ferromagnet. The homogeneous external field does not couple directly with the Néel mode which is the ground state of the AF. Therefore, AF correlations can disappear at T_N even in the presence of an external field. The presence of this field does, however, affect the magnitude of T_N.

An interesting detail concerning the spherical model antiferromagnet is discussed in the following Problem.

....

Problem 2.13. The susceptibility in (2.11.4) is the longitudinal susceptibility, as previously defined in connection with Néel's theory (Sect.2.6). Calculate the perpendicular susceptibility in the spherical model, and compare with the longitudinal susceptibility and with MFT.
....

2.12 Spherical Model Spin Glass

In recent years, a great deal of interest has centered about the spin-glass phase. Physically, it concerns magnetic ions having random interactions with one another, as in moderately dilute solutions of magnetic ions such as Mn or Fe in nonmagnetic host metals such as Cu. The RKKY interactions [2.12] are of the form $-J_{ij}S_iS_j$ between every pair of spins, with

$$J_{ij} = J_0|R_{ij}|^{-3} \cos 2k_F|R_{ij}| \qquad (2.12.1)$$

and as the position of the individual spins is a random variable, this long-ranged interaction (in the sense (b) of Sect.2.7) is oscillatory in sign and random in magnitude. In other prototype spin glasses, the interactions are short-ranged but the presence or absence of spins on the prescribed sites introduces a random variable into the analysis.

In any given model, one must know the particular geometry of the spin arrangement (e.g., randomly distributed spins, or clustered, or on a regular array) and of the bonds (e.g., nearest-neighbor ferromagnetic, or n.n. AF, or long ranged as in (2.12.1) and of the external fields (e.g., random external field, or homogeneous of given magnitude, etc.). The free energy is calculated as -kT ln Z for the given configuration of random variables. (We define a configuration of random variables as the statement about spin lattice geometry, the sign and magnitude of the individual J_{ij} and the external field, all subject to given probability distributions.) As the free energy is extensive, in the limit $N \to \infty$ all possible configurations will be experienced. Thus, the desired object is the free energy per spin in the thermodynamic limit:

$$f = F/N = -\frac{kT}{N} <\log Z> \qquad (2.12.2)$$

where Z is the trace over spins, for the given random configuration, and < > signifies average over all random configurations.

If the random configurations were annealed instead of quenched, we could follow the *much* simpler procedure of averaging Z over random configurations. Unfortunately, the activation energy to move magnetic ions in the metal, or to vary any of the other random variables, is generally so large that in the range of temperatures of interest (0 K-10^3K) these random variables may be considered frozen-in, i.e., quenched.

The last step, the average over random configurations in (2.12.2), is the hardest. Fortunately there exist examples in which all random configurations (excepting a set of measure zero) yield the same Z in the thermodynamic limit. Because of this feature, they are easily solved. The one we study in this section is a version of the spherical model spin glass first solved by Kosterlitz et al. [2.20]. But before specializing to their version of long-ranged forces, let us develop the formalism for arbitrary spherical models. Incorporating the constraint explicitly, one writes an Hamiltonian:

$$H = -\frac{1}{2} \sum_i \sum_{j \neq i} J_{ij} S_i S_j + \mu \left(\sum_i S_i^2 - N \right) . \tag{2.12.3}$$

It must be noted that, in general, H depends on the particular configuration or set of $\{J_{ij}\}$ which is quenched.

Proceeding, we diagonalize H so as to evaluate Z in convenient form. Label the eigenvalues and orthonormal eigenvectors by a set of N quantum numbers $\{\alpha\}$; thus a typical eigenvalue equation is

$$-\sum_j J_{ij} \phi_\alpha(j) = 2E_\alpha \phi_\alpha(i) \tag{2.12.4}$$

taking the eigenvalue to be $2E_\alpha$ for future convenience. Now H takes on the following form

$$H = \sum_\alpha (E_\alpha + \mu) X_\alpha^2 - N\mu \tag{2.12.5}$$

in terms of the normal mode amplitudes X_α. The individual spins are now given by

$$S_i = \sum_\alpha X_\alpha \phi_\alpha(i) \tag{2.12.6}$$

and as the $\phi_\alpha(i)$ are an orthonormal set, the conditions $\langle \sum_i S_i^2 \rangle = N$ becomes

$$\left\langle \sum_\alpha X_\alpha^2 \right\rangle = N , \tag{2.12.7}$$

exactly equivalent to $\partial f/\partial \mu = 0$, after traces and configurational averages are performed.

Along with the set of eigenvalues $\{E_\alpha\}$ one introduces the concept of a density of states $\rho(E)$,

$$\rho(E) \equiv \sum_\alpha \delta(E - E_\alpha) \tag{2.12.8}$$

and of a local density of states at the spin S_i,

$$\rho_i(E) = \sum_\alpha \phi_\alpha^2(i) \delta(E - E_\alpha) , \tag{2.12.9}$$

clearly,

54

$$\sum_i \rho_i(E) = \rho(E) \quad .$$

The free energy in the spin glass is obtained in a manner similar to (2.9.14),

$$F_{SG}/N = -\frac{1}{2} kT \ln 2\pi - \mu + \frac{1}{2} kT \frac{1}{N} \sum_\alpha \ln \frac{2}{kT} (\mu + E_\alpha)$$

$$= -\frac{1}{2} kT \ln 2\pi - \mu + \frac{1}{2} kT \frac{1}{N} \int dE\rho(E) \ln \frac{2}{kT} (\mu + E) \quad ,$$

but must now be averaged over random configurations:

$$f = -\frac{1}{2} kT \ln 2\pi - <\mu> + \frac{1}{2} kT \frac{1}{N} \int dE <\rho(E) \ln \frac{2}{kT} (\mu + E)> \quad . \tag{2.12.10}$$

This serves to define $<\rho(E)>$, the configuration-averaged density of states, and shows that a knowledge of the thermodynamic functions requires only a knowledge of the statistical distribution of the eigenvalues of J_{ij}.

At this point, let us specialize to the spin glass of [2.20]. The bonds J_{ij} ($i \neq j$) are random, $<J_{ij}> = 0$ and $<J_{ij}^2> = J^2/N$, J being the strength of the random bonds.

The distribution of eigenvalues of such a matrix is known precisely in the thermodynamic limit [2.21,22]; it is Wigner's semicircular density of states. For each spin S_i and configuration,

$$\rho_i(E) = \frac{2}{\pi J} \left[1 - (E/J)^2\right]^{\frac{1}{2}} \quad , \quad \text{each } \mu = <\mu> \quad , \tag{2.12.11}$$

and, as this result is explicitly independent of i, the configuration-averaged density of states is simply

$$<\rho(E)> = \frac{2N}{\pi J} [1 - (E/J)^2]^{\frac{1}{2}} \quad . \tag{2.12.12}$$

Substitution of this formula into (2.12.10) yields the free energy in this model. The spherical constraint equation obtained by differentiating f, $\partial f/\partial\mu = 0$, has the following aspect:

$$\frac{kT}{\mu J} \int_{-J}^{+J} dE[1 - (E/J)^2]^{\frac{1}{2}} \frac{1}{\mu + E} = \frac{kT}{J} [\tau - (\tau^2 - 1)^{\frac{1}{2}}] = 1 \tag{2.12.13}$$

in terms of the natural variable $\tau \equiv \mu/J$. Defining $kT_c = J$, this equation yields

$$\tau = \frac{T^2 + T_c^2}{2TT_c} \quad \text{for} \quad T \geqslant T_c \quad , \tag{2.12.14}$$

with τ sticking to its value $\tau_c = 1$ in the range $T \leqslant T_c$. Comparison with the ferromagnetic spherical model, (2.9.22 and 23), shows how to obtain the specific heat. One can easily express it in closed form:

$$c = \begin{cases} \frac{1}{2} k & (T \leqslant T_c) \\[2ex] \frac{1}{2} kT_c^2/T^2 & (T > T_c) \end{cases} \qquad (2.12.15)$$

While the specific heat is continuous at T_c, its derivative is not. According to the conventional classification scheme, the phase transition at T_c is of *third* order, as the leading discontinuity occurs in a third derivative of the free energy. But this point may well be academic, for it is virtually impossible to detect experimentally discontinuities in third order or higher, therefore the existence or nonexistence of a phase transition in this model may not be of any consequence. Indeed, we observe in the following section that any external field will eliminate the phase transition altogether.

In many important respects, the properties of this long-ranged (all spins interact with one another) random model mimic those of the three-dimensional ferromagnet studied previously. This is not a coincidence. If we identify J here with 3J in the ferromagnet, we find that the respective *dos* functions have similar features, shown in Fig.2.15 (some details are left to the reader in Problem 2.14). As the free energy is determined entirely by the *dos*, the thermodynamic resemblence follows directly. Unfortunately, such nice mappings do not carry over into Ising and quantum-mechanical models of magnetism.

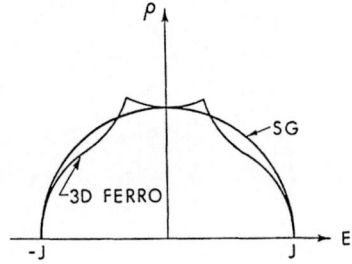

Fig.2.15. Wigner's semicircular density of states $\rho(E)$ of a perfectly random matrix J_{ij}, as derived in Sect.2.13. It is marked SG, and contrasted to the density of states for a nearest-neighbor sc lattice (Sect.2.9)

. . . .

Problem 2.14. Calculate the *dos* function for the 3D nn ferromagnet:

$$\rho(E) = \sum_{\mathbf{k}} \delta[E + (J/3)(\cos k_x + \cos k_y + \cos k_z)] \quad , \qquad (2.12.16)$$

and show that near the band edges, $E = \pm J$, it shares the 1/2 power law behavior with Wigner's semicircular *dos*, (2.12.12). Show the integrals over (2.12.12 and 16) are both equal to N. Discuss (2.12.16) at $E = \pm J/3$, the sites of Van Hove singularities (see any recent book on solid-state theory).
....

2.13 Magnetic Properties of Spin Glass

As the τ-parameter sticks to its value $\tau_c = 1$ below T_c, there must occur a condensation into the lowest mode at $E = -J$. The similarity of the random model with the 3D ferromagnet ceases if we examine the nature of the condensed phase. In the latter, it is the macroscopic magnetization that grows as the temperature is lowered. In the case of the random interactions, the corresponding normal mode varies from configuration to configuration and is a random variable. So, although there exists an internal order variable which describes the freezing out of the spin glass at low temperature — the amplitude X_0 of the condensed phase — satisfying an equation of the type (2.9.21),

$$\frac{T}{T_c} + (X_0^2/N) = 1 \qquad (T \leqslant T_c) \quad , \tag{2.13.1}$$

the corresponding normal mode is unknowable. It is therefore that much more surprising that applying an external magnetic field to this system eliminates the phase transition, and along with it, the macroscopic condensation.

In studying the magnetic response of the random model, we can even generalize it a bit. Let each J_{ij} be a random variable, such that

$$\langle J_{ij} \rangle = J_0/N \quad \text{and} \quad \langle J_{ij}^2 \rangle = J^2/N \quad . \tag{2.13.2}$$

Positive J_0 signifies a tendency to ferromagnetism, negative J_0 antiferromagnetism. This model combines the long-ranged systematic interactions examined in Sect.2.7 with the random ones. We recall that the example in Sect. 2.7 involved only the $k = 0$ plane wave. There, plane waves were normal modes. Here, we can use a procedure known as tridiagonalization [2.21] to obtain the spectral decomposition of each plane wave.

First, we write the Hamiltonian, including the interaction with the external field, and the spherical condition in terms of plane wave states — which we call spin wave states henceforth.

$$H = H_0 + H' \quad , \tag{2.13.3}$$

with

$$H_0 = -\frac{1}{2} \sum_{kk'} W_{kk'}\sigma_k\sigma_{k'} + \left(\sum_k |\sigma_k|^2 - N\right)\mu \tag{2.13.4}$$

and

$$H' = -BN^{\frac{1}{2}}\sigma_0 \tag{2.13.5}$$

using the definition of a spin wave:

$$\sigma_k = N^{-\frac{1}{2}} \sum_i \exp(ik \cdot R_i)S_i \quad . \tag{2.13.6}$$

The spin wave states have an expansion in normal modes $\psi_\alpha(k)$:

$$\sigma_k = \sum_\alpha X_\alpha \psi_\alpha(k) \quad . \tag{2.13.7}$$

In terms of the amplitudes X, the various quantities above take on the following aspect when reduced to normal modes:

$$H_0 = \sum_\alpha (E_\alpha + \mu)X_\alpha^2 - N\mu \tag{2.13.8}$$

and

$$H' = -BN^{\frac{1}{2}} \sum_\alpha X_\alpha \psi_\alpha(0) \quad . \tag{2.13.9}$$

The spherical condition (2.12.7) remains unaffected.

We now carry out the decomposition of each spin wave into normal modes to implement (2.13.7). First, consider the k = 0 wave in (2.13.6):

$$\sigma_0 = N^{-\frac{1}{2}} \begin{bmatrix} 1 \\ 1 \\ \vdots \\ 1 \end{bmatrix} \quad . \tag{2.13.10}$$

We will denote it $\mathbf{v}_0$, and study the effects of the J_{ij} matrix on it.

$$-\mathbf{J} \cdot \mathbf{v}_0 \equiv -\sum_j J_{ij}v_0(j) = m_{00}v_0(i) + m_{01}v_1(i) \quad . \tag{2.13.11}$$

The individual J_{ij}'s (i ≠ j) are random, subject to the constraints (2.13.2) on their average and on their square. The various quantities are obtained as follows:

$$m_{00} = -\sum_{i,j} v_0(i)J_{ij}v_0(j) \tag{2.13.12}$$

assuming $\mathbf{v}_1$ is a new vector, orthogonal to $\mathbf{v}_0$ and like it, of unit length. It follows that

$$m_{01}^2 = \mathbf{v}_0 \cdot (\mathbf{J} + m_{00}\mathbf{1})^2 \cdot \mathbf{v}_0 \tag{2.13.13}$$

and

$$v_1 = m_{01}^{-1}(- J - m_{00}1) \cdot v_0 \quad . \tag{2.13.14}$$

Proceeding, one allows J to act on v_1, generating a new vector v_2 in addition to the known v_1 and v_0. Continuing this procedure, one obtains a matrix $m_{n,n'}$ which vanishes unless $n = n'$ or $n = n' \pm 1$ (a tridiagonal matrix). In the particular case at hand, even though the original J_{ij} are random, the $m_{n,n'}$ tridiagonal matrix becomes sharp in the thermodynamic limit, having the values [2.21]:

$$m_{0,0} = -J_0 \quad , \quad m_{n,n} = 0 \quad (n \neq 0) \quad , \quad \text{and}$$

$$m_{n,n+1} = -J = m_{n+1,n} \quad , \quad \text{all} \quad n \geq 0 \quad . \tag{2.13.15}$$

We have analyzed precisely such a matrix in [Ref.2.1, pp.274-275]. The conclusions were: there is a continuous spectrum of eigenvalues $2E$ ranging from $-J < E < +J$ and, in addition, a bound state *below* the continuum if $J_0 > |J|$, and above the continuum if $J_0 < -|J|$. The density of states which describes the continuum is:

$$\rho_0(E) = \frac{2}{\pi} [1 - (E/J)^2]^{\frac{1}{2}} \frac{J}{J_0^2 + J^2 + 2J_0 E} \tag{2.13.16}$$

labeling it with subscript 0 to indicate the $k = 0$ sector, and assuming $|J_0| < |J|$ so there are no bound states. Turning to the $k \neq 0$ sectors, we start with σ_k as v_0 replacing (2.13.10), and follow the same procedure. The results are similar, except that J_0 does not appear in m_{00}. Thus, each sector $k \neq 0$ has a *dos*:

$$\rho_k(E) = \frac{2}{\pi J} [1 - (E/J)^2]^{\frac{1}{2}} \quad . \tag{2.13.17}$$

The *combined* weight of the $k \neq 0$ sectors is N-1. Thus, unless something special occurs, the $k = 0$ *dos* has zero statistical weight in the thermodynamic limit and can be ignored.

The special cases where the $k = 0$ sector contributes are of two distinct types. First, if $J_0 > |J|$, a bound state develops at eigenvalue $E_0 < -|J|$. As the temperature is lowered and μ becomes smaller, there will come a temperature where $\mu - |E_0|$ will vanish. This condensation into the $k = 0$ sector occurs before any of the integrals become singular, and thus dominates the thermodynamics even though the statistical weight of the bound state is negligible. As the condensed state is not orthogonal to σ_0, there is partial ferromagnetism —a form of "mictomagnetism" distinct from the spin glass phase (insofar as *spontaneous* magnetization will be found at sufficiently low temperature). The critical temperature will be a function of J_0. Some properties of this phase were treated in [2.20]; we ignore it henceforth.

The second special case just involves the application of an external magnetic field. The spin glass is quite sensitive to an applied field even though its paramagnetic susceptibility is finite. We shall return to this at length. But first, we may inquire as to the behavior of the model when $J_0 < -|J|$, i.e., when a bound state appears in the $k = 0$ sector *above* the continuum. In that case the bound state factor $(\mu + E_0)$ never vanishes, and the bound state never contributes due to the negligible statistical factor $1/N$. This lack of symmetry between negative and positive bonds was previously noted in the long-range model (Sect.2.7). The *spin glass phase* is the term given to the stable thermodynamic condensed phase which is found for J_0 in the range $J_0 < |J|$; for example, T_c will be independent of J_0 *for any J_0 in this range*, and so is the zero-field specific heat, which continues to be given by (2.12.15) precisely!

We now examine the effects of an applied external field B.

The magnetization in the spin glass phase (*no* bound state) is:

$$m = B \frac{K_1}{\pi J_0} \int_{-J}^{+J} dE[1 - (E/J)^2]^{\frac{1}{2}} \frac{1}{\gamma_0 + (E/J)} \frac{1}{\mu + E} \tag{2.13.18}$$

where $\gamma_0 \equiv (J^2 + J_0^2)/(2JJ_0)$. K_1 is just a constant of proportionality permitting us to introduce convenient units — say, express m in units of the saturation magnetization, and B in such units that we recover $\chi = \mathbb{C}/T$ at high temperature.

The spherical condition is easily expressed in terms of the spin wave states. Only the states connecting to $k = 0$ are affected by the external field, and so we obtain a formula only slightly more complex than (2.12.13):

$$\frac{kT}{J} [\tau - (\tau^2 - 1)^{\frac{1}{2}}] + (BK_2)^2 \frac{1}{\pi J_0} \int_{-J}^{+J} dE[1 - (E/J)^2]^{\frac{1}{2}}$$

$$\times \frac{1}{\gamma_0 + (E/J)} \frac{1}{(\mu + E)^2} = 1 \tag{2.13.19}$$

with K_2 determined by the choice of units. Note that J_0 enters this expression *only* when there is an external field. Performing the integrations and adjusting the constants K_1 and K_2 suitably, we obtain for the magnetization,

$$m = (B\mathbb{C}/2T_c)(J/J_0) \left[\frac{\tau - (\tau^2 - 1)^{\frac{1}{2}} - \gamma_0 + (\gamma_0^2 - 1)^{\frac{1}{2}}}{\gamma_0 - \tau} \right] \tag{2.13.20}$$

with $kT_c = J$ as before. This spherical condition is

$$\frac{kT}{J}[\tau - (\tau^2 - 1)^{1/2}] + (B\mathbb{C}/2T_c)^2(J/J_0)\times$$

$$\left\{(\gamma_0 - \tau)^{-2}[\gamma_0 - (\gamma_0^2 - 1)^{1/2} - \tau + (\tau^2 - 1)^{1/2}] + (\gamma_0 - \tau)^{-1}\right.$$

$$\left.\times \left[\frac{\tau}{(\tau^2 - 1)^{1/2}} - 1\right]\right\} = 1 \quad . \tag{2.13.21}$$

These equations simplify considerably if we introduce two quantities with the units of temperature, $T_F = 2J_0/k$ and $\theta = T_c[\tau - (\tau^2 - 1)^{1/2}]^{-1}$. The restriction $J_0 \leqslant |J|$ is equivalent to the inequalities,

$$\theta \geqslant T_c \geqslant T_F \quad . \tag{2.13.22}$$

If J_0 is negative (AF), so is T_F. The following relations are easily demonstrated:

$$\gamma_0 = \frac{T_F^2 + T_c^2}{2T_F T_c} \quad \text{and} \quad \tau = \frac{\theta^2 + T_c^2}{2\theta T_c} \quad . \tag{2.13.23}$$

It is then a matter of some algebraic manipulations to simplify the formula for m to the familiar form,

$$m = \frac{B\mathbb{C}}{\theta - T_F} \tag{2.13.24}$$

and the spherical condition to polynomial form,

$$\frac{T}{\theta} + \frac{(B\mathbb{C}\theta)^2}{(\theta - T_F)^2(\theta^2 - T_c^2)} = 1 \quad . \tag{2.13.25}$$

Specifying the parameters T_c and T_F, and fixing B, one can vary θ in the two equations above, obtaining $T(\theta)$ and $m(\theta)$, and thereby $m(T)$. For *any finite B* (however small) *there is no phase transition*, as the solutions are analytic in θ. As $B \rightarrow 0$, however, a cusp develops in the function $\theta(T)$, i.e., a discontinuity in $T(\theta)$. Thus, in the $B = 0$ limit there is the previously mentioned third-order phase transition with the specific heat function still given precisely by (2.12.15), independent of T_F. The zero-field susceptibility *is* sensitive to T_F. Calculating the ratio $(m/B\mathbb{C})$ in the $B = 0$ limit we obtain

$$\chi_0 = \frac{\mathbb{C}}{T - T_F} \quad \text{for} \quad T > T_c \quad , \quad \text{and} \quad \chi_0 = \frac{\mathbb{C}}{T_c - T_F} \quad , \quad T \leqslant T_c \quad .$$

$$\tag{2.13.26}$$

Combining (2.13.24 and 25) yields

$$m = \left\{[1 - (T_c/\theta)^2][1 - (T/\theta)]\right\}^{\frac{1}{2}} \qquad (2.13.27)$$

for arbitrary B,T. Because θ increases monotonically with B, this expression proves that the *maximum* m is 1, the saturation magnetization. This, and the Curie-Weiss result (2.13.26), allow convenient comparison with experiment and with other theories. The most recent experiments on the spin-glass phase are displayed in Fig.2.16.

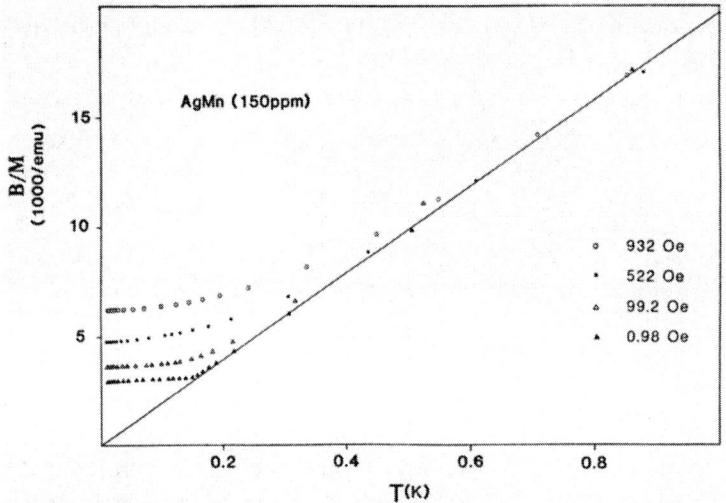

Fig.2.16. B/M as function of T for various B in a spin glass consisting of dilute solid solution of Mn in Ag, as measured by M. Novak, O.G. Symko, d. Zheng: Private communication (1984). The dependence on B tracks the theoretical solution of (2.13.24,25), with T_F = order of $10^{-3}K \approx 0$, and indicates $T_c \approx 0.15$ K . These experiments show a large discrepancy with the spherical model theory in *strong* magnetic fields, and indicate the need for a more rigorous theory

Other interesting or nontrivial applications of the spherical model are listed in [2.23].

In concluding this section on the spherical model spin glass, it may be appropriate to note the very great difficulties that were faced in the more realistic Ising or vector spin glasses. The conditions $S_i^2 = 1$ for N spins are difficult to preserve alongside the stochastic features of the problem, and no satisfactory resolution exists at the data of writing. The impetus for much of the present research came in a paper by *Sherrington* and *Kirkpatrick* [2.24], who introduced the long-range model defined by (2.13.2), treating the spins in the Ising manner, $S_i = 1$. Other popular realistic models include the nearest-neighbor random bond model of *Edwards* and *Anderson* [2.25] and the treatment of the realistic indirect exchange interactions in metals by *Walker* and *Walstedt* [2.26], perforce numerical. But the mean-field aspect of the Sherrington-Kirkpatrick model made it the most attractive candidate for a universal theory, in the sense of the Weiss-Néel theories, and it has therefore been tested and studied from many different points of view.

The results of such studies have been most confusing, leading *Kosterlitz* et al. to their study [2.20] of the spherical version of the Sherrington-Kirkpatrick (S-K) model, which we have outlined in these sections. The lack of ergodicity of the S-K model has been found [2.27] to be related to *relaxation times* $O(N^{\frac{1}{2}})$ instead of the usual $O(1)$.

Parisi has introduced a functional order parameter [2.28], while *Sompolinsky* [2.29] has a time-dependent variant thereof, and *Jonsson* [2.30a] presented an order parameter which is a function of two variables, following the suggestion of *de Dominicis* et al. [2.30b].

The difference between the simple spherical model solution we have reproduced here, and the more realistic models in the literature, lies in the activation barriers which prevent individual spins in the latter from flipping (from up to down or vice versa) as the temperature is changed or the magnetic field is applied. The failure of linear response theory which results was first noted by *Parisi* [2.28], and by *Bray* and *Moore* [2.31a] who where studying the distribution of relaxation times. A numerical experiment conducted by *Bantilan* and *Palmer* [2.31b] casts some light on this situation. Studying the ground state ($T = 0$) of 200 Ising spins, averaged over 50 different sample distributions of random J_{ij}'s, they found *two* distinct curves for $m(B)$. The lower one exhibits $\chi = \partial m/\partial B = 0$ at small B, and is the result of varying the ground state continuously with applied field. The upper curve, displaying finite susceptibility, is the result of optimizing the ground state ab initio at each value of the applied field. The latter is easily obtained in the spherical model approximation, but the former may simulate the experimental situation more accurately.

Due to the many uncertainties at present, we urge the interested reader to consult the most recent literature prior to pursuing this rapidly evolving subject [2.32].

....
Problem 2.15. *Prove* that only the *maximum* calculated susceptibility is
thermodynamically stable. As a consequence, the lower of the curves obtained
by Bantilan and Palmer must be thermodynamically unstable, although the relaxation time might be ∞ in the thermodynamic limit.
....

2.14 Thermodynamics of Magnons

Magnons are the quantized elementary excitations in magnetic systems with
continuous symmetry, including both the Heisenberg and the itinerant electron models. In the classical limit they are the spin waves. But at low
temperatures, where quantum effects are important, it is the study of the
magnons that reveals the correct thermodynamic properties.

We have previously observed the mapping of magnon states onto harmonic
oscillators [Ref.2.1, Chap.5] and observed that at low density, their commutation relations are those of photons, phonons, ... etc., i.e., that of
bosons. Starting with a ferromagnetic state of all spins up we introduce a
small number of magnons, say $\mathcal{N}$. As each decreases the total magnetization
by one unit of spin angular momentum ($\hbar = 1$), the total magnetization is now

$$\mathcal{M}_z = Ns - \mathcal{N} = \mathcal{M}_0 - \mathcal{N} \tag{2.14.1}$$

where the product Ns represents the saturation magnetization of N spins of
length s, also denoted $\mathcal{M}_0$. If the magnons are present owing to thermal
fluctuations appropriate to temperature T, we can use (2.8.6) to obtain $\mathcal{M}_z$
as a function of T. We denote each normal mode by $\mathbf{q}$:

$$\mathcal{M}_z(T) = \mathcal{M}_0 - \sum_{\mathbf{q}} \left[e^{\omega(\mathbf{q})/kT} - 1 \right]^{-1} \quad . \tag{2.14.2}$$

Such sums are best expressed in terms of the normal mode *dos*, $\rho(\omega)$, which
incorporates the dispersion (dependence of ω on $\mathbf{q}$) and dimensionality both.
With the substitution

$$\sum_{\mathbf{q}} \to N \int d\omega \rho(\omega) \quad ,$$

we find that the magnetization is

$$\mathcal{M}_z(T) = \mathcal{M}_0 \left[1 - \frac{1}{s} \int d\omega \rho(\omega)(e^{\beta\omega} - 1)^{-1} \right] \quad . \tag{2.14.3}$$

For an ordinary ferromagnet in 3D, $\omega \propto \mathbf{q}^2$ and therefore $\rho(\omega) \propto \omega^{\frac{1}{2}}$. Assuming
that the exponential factor cuts off the integral sufficiently that we may

take the upper limit to ∞, we make the substitution of variables $\beta\omega = x$ and with

$$\int d\omega\rho(\omega)(e^{\beta\omega} - 1)^{-1} = (kT)^{3/2} \int dx\rho(x)(e^x - 1)^{-1}$$

factor the temperature dependence from the integral. The result is

$$\mathcal{M}_z(T) = \mathcal{M}_0\left[1 - (T/T_0)^{3/2}\right] \qquad (2.14.4)$$

the celebrated "$T^{3/2}$ law" of F. Bloch, with T_0 a lumped constant involving the integral and s. If we define the Curie temperature by the vanishing of spontaneous long-range order (LRO), $\mathcal{M}_z$ in this case, then it is T_0. But there is reason to believe that the $T^{3/2}$ law fails near T_c owing to our neglect of a number of mathematically and physically important concerns. Figure 2.17 shows in the case of some Gd alloys that this law is substantially satisfied until T reaches approximately 2/3 of the critical temperature, i.e., is quite satisfactory at low temperatures. At finite temperatures there are two sources of error: a change in dispersion relations at finite q (including Brillouin zone edge effects), and the interactions and scattering of magnons with one another denoted generically as nonlinear effects. The higher the temperature, the larger the q's which are excited, so the greater the first correction, but the greater the number of magnons, the greater the second as well.

Consider an Heisenberg ferromagnet with nearest-neighbor bonds on a sc lattice, spin magnitudes s ≫ 1, so that expansions in powers of 1/s are justified [Ref.2.1, Sect.5.7]. The magnetization, in units of $\mathcal{M}_0$, is

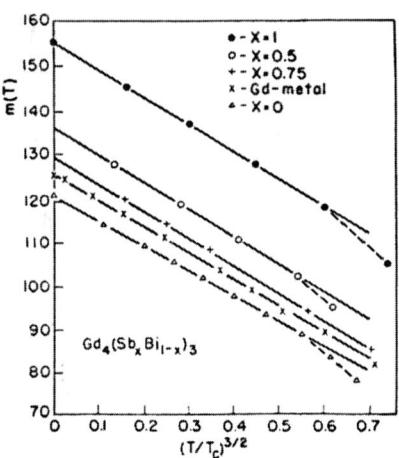

Fig.2.17. Verification of Bloch's $T^{3/2}$ law in several gadolinium (s = 7/2) alloy ferromagnets [2.34]. Slight downward deviations near T_c may be due to magnon interactions: cf. Fig.2.18

$$m(T) = 1 - s^{-1}(2\pi)^{-3} \int\int\int_{-\pi}^{+\pi} dk_x dk_y dk_z (e^{\beta\omega(\mathbf{k})} - 1)^{-1} \qquad (2.14.5)$$

where

$$\omega(\mathbf{k}) = 2sJ(3 - \cos k_x - \cos k_y - \cos k_z) \qquad (2.14.6)$$

reflects the lattice structure; it is approximated by sJk^2 only for $k^2 \ll 1$. We can make use of the properties of Bessel functions of imaginary argument I_p.

$$I_p(z) = \frac{1}{2\pi} \int_{-\pi}^{\pi} d\theta \; e^{z \cos\theta} \cos p\theta = \left(\frac{z}{2}\right)^p \sum_{m=0}^{\infty} \frac{(z/2)^{2m}}{m!(|p| + m)!} \quad . \qquad (2.14.7)$$

Expanding the denominator of (2.14.5) in a geometric series in the exponential, we see that only I_0 enters into the expansion, which takes the following aspect:

$$m(T) = 1 - \frac{1}{s} \sum_{n=1}^{\infty} \left[e^{-ng} I_0(ng) \right]^3 \qquad (2.14.8)$$

where $g = 2sJ/kT$. The bessel functions have asymptotic expansions which are convenient at large arguments, low temperature in this instance, i.e.,

$$I_0(z) \sim \frac{e^z}{(2\pi z)^{\frac{1}{2}}} \left[1 + \sum_{r=1}^{\infty} \frac{1^2 \times 3^2 \cdots (2r - 1)^2}{r! 2^{3r} z^r} \right]$$

$$= \frac{e^z}{(2\pi z)^{\frac{1}{2}}} \sum_{r=0}^{\infty} \frac{[\Gamma(r + \frac{1}{2})]^2}{\pi r! (2z)^r} \quad . \qquad (2.14.9)$$

Inserted into the preceding, this both yields the series:

$$m(T) = 1 - B_{3/2}(T/T_0)^{3/2} - B_{5/2}(T/T_0)^{5/2} - B_{7/2}(T/T_0)^{7/2} - \cdots \qquad (2.14.10)$$

and the values of the coefficients $B_{n/2}$, which we omit (in an external field the B's are additionally functions of the applied field; for details see [2.33]).

The asymptotic expansion is no longer accurate when kT is of the order of sJ, but at these temperatures the distribution becomes quasi-classical. Expanding the Bose-Einstein function in leading powers of sJ/kT, we have

$$(e^{\omega/kT} - 1)^{-1} = \frac{kT}{\omega} - \frac{1}{2} + O(\omega/kT) \qquad (2.14.11)$$

retaining only leading terms. Over the range of temperatures $T \gtrsim T_c s^{-1}$ we then calculate

$$m(T) \cong 1 + \frac{1}{2s} - \frac{kT}{2Js^2}\left(\frac{1}{2\pi}\right)^3 \int_{-\pi}^{\pi} \frac{dk_x dk_y dk_z}{3 - \cos k_x - \cos k_y - \cos k_z}$$

$$\cong 1 + \frac{1}{2s} - \frac{kT}{6Js^2} W \qquad\qquad (2.14.12)$$

where the Watson integral $W = 1.516 \ldots$, as given in Sect.2.9. Calculating the Curie temperature T_c by setting $m(T_c) = 0$, we obtain

$$kT_c = 3.96Js^2\left(1 + \frac{1}{2s}\right) \quad s \gg 1 \quad . \qquad\qquad (2.14.13)$$

For comparison, the same lattice may be treated in the molecular field approximation, where each spin interacts with a molecular field,

$$H = -S_i \cdot B_m$$

made up of the average force exerted by its neighbors,

$$B_m = 2J \left\langle \sum_\delta S_\delta \right\rangle_{TA} \quad .$$

One may calculate $<S_i>_{TA}$ in the standard way, and then solve the molecular field equation,

$$<S_i>_{TA} = <S_j>_{TA} \quad \text{for all } i,j \quad .$$

The temperature at which this constitutive equation ceases to have only a non-trivial solution is the Curie-Weiss temperature $T_c = \theta$, the temperature at which $\chi = \mathbb{C}/(T - \theta) \to \infty$:

$$k\theta = 4Js^2\left(1 + \frac{1}{s}\right) \quad . \qquad\qquad (2.14.14)$$

The agreement of this high-temperature theory with the low-temperature result is truly astounding. But very accurate high-temperature series extrapolation methods have shown both estimates to be some 30% too high!

Taking into account the *real* shifts in magnon energies due to magnon-magnon interactions overcorrects this situation near T_c, as we shall now observe. We recall [Ref.2.1, Sect.5.7] that the energy of a magnon depends on the occupancy of all the other magnons, according to the formula:

$$\varepsilon_k = \hbar\omega_k - \frac{1}{Ns} \sum_{k'} (\hbar\omega_k + \hbar\omega_{k'} - \hbar\omega_{k-k'} - \hbar\omega_0)<n_{k'}> \quad . \qquad (2.14.15)$$

The occupation-numbers $<n_k>$ are, on thermal average,

$$<n_k> = [\exp(\varepsilon_k/kT) - 1]^{-1} \quad . \qquad\qquad (2.14.16)$$

These equations are coupled, nonlinear equations for the self-consistent determination of the nonlinear magnon spectrum. Equation (2.14.15) can be simplified somewhat if we take advantage of the cubic symmetry, and reduces to

$$\varepsilon_k = \omega(\mathbf{k})[1 - b(T)] \qquad \text{with} \qquad (2.14.17)$$

$$b(T) = (2Js^2)^{-1} \frac{1}{N} \sum_{\mathbf{k}'} \omega(\mathbf{k}')\langle n_{\mathbf{k}'}\rangle \qquad (2.14.18)$$

a function which must be calculated self-consistently. Solutions of these equations lead to several interesting results. At low temperature, the first correction to the expansion in (2.14.10) occurs via a term of the form

$$-C(T/T_0)^{8/2} \qquad (2.14.19)$$

and is practically negligible. Near T_c, the corrections are similar to the experimental data (Fig.2.17), but even more severe, as shown in Fig.2.18. The consequence is that the theory predicts a first-order phase transition with a small jump in m(T) at T_c. The calculated value of T_c is now within a few percent of the exact value, but the jump in m(T) at T_c is incorrect. This unphysical jump is probably the result of our total neglect of lifetime broadening effects, i.e., of magnon-magnon scattering. A theory which dealt properly with this problem in a simple manner would be desirable, but none is available.

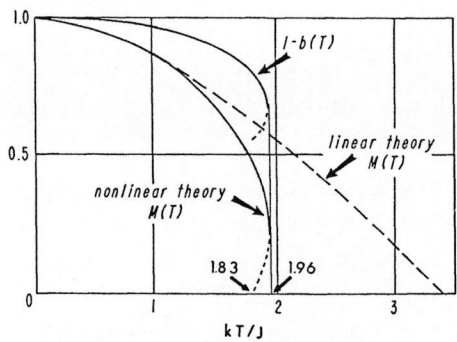

Fig.2.18. Modifying magnon dispersion to include nonlinearities (self-energy modifications due to magnon-magnon interactions), (2.14.17,18), leads to improved agreement with experiment (Fig.2.17) but also to an unphysical jump in m(T) at T_c [2.35]

So far we have concentrated only on the magnetization, but the internal energy and its derivative, the heat capacity, are of equal interest. Having determined B(T) self-consistently, we can immediately write

$$U = \sum \omega(\mathbf{k})\left[1 - \frac{1}{2} b(T)\right]\langle n_{\mathbf{k}}\rangle \qquad (2.14.20)$$

for the internal energy, taking care with the factor 1/2 not to double-count the interactions. The evaluation of such expressions proceeds along the model of the calculation of m(T) above.

Had we attempted to perform analogous calculations appropriate to one- and two-dimensional ferromagnets, we would have immediately discovered that integrals such as in (2.14.5) are divergent in 2D or in any lower dimension. As we shall see in the following section, this is related to the absence of long-range order for models with continuous symmetry in low-dimensional arrays. This also entails the breakdown of spin-wave theory and its quantized magnon version, as these are the dynamical excitations from an ordered ground state. The lack of a simple framework may explain why theoreticians have been studying low-dimensional systems with such great fascination in recent years, with much new mathematical and physical understanding resulting therefrom.

....

Problem 2.16. Develop the low-temperature expansion of U (2.14.20) in powers of $T^{\frac{1}{2}}$, indicating the contribution of b(T).

....

2.15 Magnetism in Two Dimensions

In this section, we seek to reconcile two facts which are in apparent conflict: (i) there can be no long-range order in magnetic systems with continuous symmetry in 2D at any finite temperature (ii) such systems *can* have a phase transition at finite T, with such discontinuities as infinite susceptibility, etc.

Of these two observations, (i) has the greater force, being the consequence of a rigorous theorem due to *Mermin* and *Wagner* [2.36]. The second was explained by *Stanley* and *Kaplan* [2.37] in the following terms: if $\chi(T)$ diverges at, and below, T_c, the implication is that the correlation function does not decay exponentially, but rather as some power law:

$$\Gamma(R) \equiv <S_0 \cdot S_R> \sim 1/R^{\eta(T)} \tag{2.15.1}$$

with $\eta < 2$ for $T \leqslant T_c$. Then, the susceptibility diverges:

$$\chi \sim \sum \Gamma(R) = \infty$$

for $T \leqslant T_c$, while if $\Gamma(R)$ decays exponentially above T_c, the susceptibility will be finite above that temperature. It is interesting that this can occur without $\Gamma(\infty)$ being finite — i.e., without LRO. Such features must be very much model-dependent. They are not seen in the 2D spherical model. Stanley and Kaplan illustrate with the Heisenberg ferromagnet, finding near T_c:

$$\chi \sim (T - T_c)^{-\gamma_c} \qquad\qquad\qquad\qquad (2.15.2)$$

with $\gamma_c \cong 5/2 + 2/(3s^2)$ and $kT_c \cong 1/5(z-1)[2s(s+1)-1]J$ in lattices with coordination number $z > 3$. It seemed for a while that these results (based on
some dozen terms in a high-temperature series expansion) would meet the test
of time. They were experimentally confirmed by observations on a number of
2-dimensional antiferromagnets and ferromagnets [2.38]. However, it is difficult to tell whether 3-dimensional effects were not intruding into the
experiments, and "numerical experiments" appear more reliable. The most recent such studies [2.39] have apparently demonstrated that $T_c = 0$ for the
classical ($s = \infty$) limit, and thus cast doubt on the Stanley-Kaplan results at
all values of s. Nevertheless, an anisotropic version of this model (the
plane-rotator model) *does* have a phase transition of the type predicted by
Stanley and Kaplan! Its interesting properties are treated in Sects.2.16-18
immediately following.

We now demonstrate the theorem of Mermin and Wagner, extending it to
itinerant ferromagnets as well. The statement is that *no magnetic LRO* (spontaneous magnetization or, for antiferromagnets, sublattice magnetization)
can be sustained at finite T, except in finite external fields B. It is im-
plemented by an inequality:

$$|m(T)| < \frac{\text{const.}}{T^{\frac{1}{2}}} \; |\ln|B||^{-\frac{1}{2}} , \quad (2D) \quad . \qquad\qquad (2.15.3)$$

The same method establishes a somewhat stronger inequality in one dimension

$$|m(T)| < \frac{\text{const.}}{T^{2/3}} \; |B|^{1/3} \quad . \qquad\qquad\qquad (2.15.4)$$

The inequalities are compatible both with divergent zero-field susceptibilities found by Stanley and Kaplan, and with the properties of a spherical model in external field. They are unrelated to the questions concerning
the existence of phase transitions in one or two dimensions.

The proof of these inequalities starts with Bogoliubov's inequality [2.40]

$$\frac{1}{2} <\{A,A^+\}><\{[C,H],C^+\}> \geq kT|<[C,A]>|^2 \qquad\qquad (2.15.5)$$

in which A and B are arbitrary operators, H is the Hamiltonian, and C is
picked to satisfy

$$B = [C^+,H] \quad . \qquad\qquad\qquad\qquad (2.15.6)$$

[,] is the usual commutator bracket, { ,} is the usual anticommutator
bracket, and we define yet another bracket:

$$(A,B) = \sum_i \sum_{j \neq i} (i|A|j)^*(i|B|j) \frac{W_i - W_j}{E_j - E_i} \qquad (2.15.7)$$

with $W_i = Z^{-1} \exp(-\beta E_i)$ the normalized Boltzmann factor. By the property of the exponential function, the following is always obeyed:

$$\frac{1}{2} \beta(W_i + W_j) > \frac{W_i - W_j}{E_j - E_i} > 0 \qquad (2.15.8)$$

and combining results,

$$(A,A) \leq \frac{1}{2} \beta <\{A,A^+\}> \quad . \qquad (2.15.9)$$

Here, Schwartz' inequality applies in the form:

$$(A,A)(B,B) \geq |(A,B)|^2 \qquad (2.15.10)$$

and with (2.15.6) defining C, we can obtain

$$(A,B) = <[C^+,A^+]> \quad \text{and} \qquad (2.15.11)$$

$$(B,B) = <[C^+,[H,C]]> \quad , \qquad (2.15.12)$$

thus establishing Bogoliubov's inequality (2.15.5).

For the Heisenberg Hamiltonian we take

$$H = - \sum_{i,j} J(R_i - R_j)S_i \cdot S_j - |B| \sum_i S_i^z \exp(iK \cdot R_i) \qquad (2.15.13)$$

picking $K = 0$ to rule out ferromagnetism, $(\pi,\pi,\pi)/a$ to rule out antiferromagnetism, or arbitrary to rule out LRO with any wavevector whatever. We define the Fourier transforms of the various quantities:

$$S^x(k) = \sum_i \exp(-ik \cdot R_i)S_i^x \quad , \quad \text{etc.} \quad , \qquad (2.15.14)$$

and take

$$C = S^x(k) + iS^y(k) = S^+(k) \quad , \quad A = S^-(-k-K) \qquad (2.15.15)$$

and form B from (2.15.6). Inequality (2.15.5) now takes the appearance

$$\frac{1}{2} <\{S^+(k+K),S^-(-k-K)\}> \geq N^2 kT\sigma^2 \div$$

$$\left\{ \frac{1}{N} \sum_{k'} [J(k') - J(k'-k)]\left\langle S^z(-k')S^z(k') + \frac{1}{4}\{S^+(k'), S^-(-k')\}\right\rangle \right.$$

$$\left. + N|B|\sigma/2 \right\} \qquad (2.15.16)$$

with $\sigma \equiv 1/N \sum_i \exp(iK \cdot R_i)<S_i^z>$ the order parameter. The denominator on the right-hand side of this relation is positive, and is always less than

$$\frac{1}{2} N \sum_i R_i^2 |J(R_i)| s(s + 1)k^2 + \frac{1}{2} N|B\sigma| \quad . \qquad (2.15.17)$$

Replacing the denominator by this upper bound and summing both sides over wavevectors, we conclude

$$s(s + 1) > 2kT\sigma^2 \int_{BZ} \frac{d^d k}{(2\pi)^d} \left[s(s + 1) \sum_i R_i^2 |J(R_i)| k^2 + |B\sigma| \right]^{-1} \qquad (2.15.18)$$

and, replacing the BZ by the smaller sphere of radius k_0 which it contains, we perform the integrations and bound the order parameter as follows:

$$\sigma^2 < 2\pi s(s + 1)k_0^{-2}\omega_0 / [kT \ln(1 + \omega_0/|B\sigma|)] \qquad (2.15.19)$$

where

$$\omega_0 \equiv \sum_i s(s + 1)k_0^2 R_i^2 |J(R_i)| \quad . \qquad (2.15.20)$$

In the limit of vanishing field B, this reduces to the result quoted in (2.15.3). In 1D, the integration results in

$$\sigma^3 < |B| \omega_0 [s(s + 1)/2kT \tan^{-1}(\omega_0/|B\sigma|)^{\frac{1}{2}}]^2 \qquad (2.15.21)$$

instead; which in the same limit, reduces to the result quoted in (2.15.4).

The vanishing of the LRO parameter cannot be simply ascribed to the softness of the magnon dispersion $\omega(k) \propto k^2$, as it applies also to antiferromagnets in which we know $\omega(k) \propto |k|$. Nor is it limited to Heisenberg models. In the case of magnetic metals, the use of spin operators such as

$$S^+(q) = \sum_k c_{k+q\uparrow}^* c_{k\downarrow} = \sum_i b_{i\uparrow}^* b_{i\downarrow} \exp(iq \cdot R_i) \qquad (2.15.22)$$

and

$$S^z(K) = \frac{1}{2} \sum_i (b_{i\uparrow}^* b_{i\uparrow} - b_{i\downarrow}^* b_{i\downarrow}) \exp(iK \cdot R_i) \qquad (2.15.23)$$

constructed out of fermion operators ($c_{k\uparrow}$ destroys an electron in Bloch state k, spin↑, whereas $b_{i\uparrow}$ destroys an electron on site R_i with the same spin index) permits virtually the same proof to go through. The fact that the spin operators commute with the one- and two-body potentials ensures that the order parameter will be the same as for a noninteracting fermion gas, i.e., zero in vanishing field. Thus, regardless whether the electronic interactions favor parallel spin alignments or antiparallel, they can only affect the short-ranged characteristics of the material.

The literature contains several less obvious extensions of these original discoveries. *Fisher* and *Jasnow* [2.41] have extended the no LRO rule to *finite* thickness slabs (eliminating the constraint to monatomic thin films, a serious experimental limitation if it existed). *Fröhlich* and *Lieb* [2.42] and *Dyson* et al. [2.43] *proved* the existence of a phase transition at finite T_c in a variety of models with continuous symmetry in 2D [2.42] and 3D [2.43]. The decay of correlation functions at large distances has been bounded by inequalities due to Griffiths, *Simon* [2.44], and others. It appears that 2D is "delicate": seemingly innocuous changes in model parameters can cause profound changes in thermodynamics.

2.16 The XY Model:1D

The peculiarities of low dimensions are nowhere better exemplified than in the XY model, the Hamiltonian of which is

$$H = -\frac{1}{2} J \sum_{(ij)} (S_i^+ S_j^- + H.C.) - B \cdot \sum_i S_i \quad . \tag{2.16.1}$$

In 1D it is exactly solvable in the extreme quantum limit of spins $s = 1/2$ (with the external field in the z-direction) by means of the Jordan-Wigner transformation to fermions. We have discussed this extensively in [Ref.2.1, Sect.5.9], and return to it in the present volume in connection with the two-dimensional Ising model in Sect.3.6. In brief, we introduce a set of fermion operators a_i^* and a_i, and their Fourier transforms c_k^* and c_k, such that the spins are expressible in terms of the a's:

$$S_i^+ = a_i^*(-1)^{Q_i} \quad , \quad \text{where} \tag{2.16.2}$$

$$Q_i = \sum_{j<i} a_j^* a_j \quad ,$$

and similarly for the Hermitean conjugate S_i^-, with

$$S_i^z = a_i^* a_i - \frac{1}{2} \quad . \tag{2.16.3}$$

The phase factors cancel if (i,j) are nearest neighbors, so for an external field in the z-direction,

$$H = -\frac{1}{2} J \sum_n (a_n^* a_{n+1} + H.C.) - B \sum_n (a_n^* a_n - \frac{1}{2}) \quad . \tag{2.16.4}$$

This quadratic form in fermion operators is diagonalized by a transformation to running waves:

$$H = \sum_k e(k) c_k^* c_k + \frac{1}{2} BN \quad , \tag{2.16.5}$$

where k runs over N equally spaced points in the interval $-\pi$, $+\pi$, and

$$e(k) = -(J \cos k + B) \tag{2.16.6}$$

is the energy of the individual fermion states. *Fadeev* and *Takhtajan* [2.45]
have remarked, in connection with the s = 1/2 Heisenberg antiferromagnet in
1D, that the appropriate excitations are fermions ($\hbar/2$) rather than magnons
($\hbar$). If we consider the XY model as an extreme limiting case of the anisotro-
pic Heisenberg model within the range $|J_z| < |J|$, then the above is a con-
firmation of their remark. (Undoubtedly the fermions form *bound pairs* of
spin $\hbar$ for $|J_z| > J$, so that the usual notions of elementary excitations
as spin-flips or magnons would be once again applicable.) Even for a classical (s →∞) isotropic Heisenberg chain, *Nakamura* and *Sasada* [2.46] again
claim that the elementary excitations are *fermionic* kinks.

The thermodynamics of the Hamiltonian (2.16.5) is obtained by using the
fermion rules (2.8.1-3). The internal energy is

$$U = \sum_k e(k) \langle n_k \rangle \quad , \quad \text{where}$$

$$\langle n_k \rangle = 1/(e^{\beta e(k)} + 1) \tag{2.16.7}$$

and dU/dT yields the heat capacity. For small J_z we can solve the thermal
Hartree-Fock equations of the anisotropic model. To H we add the perturbation

$$H' = -J_z \sum_n \left(a_n^* a_n - \frac{1}{2} \right)\left(a_{n+1}^* a_{n+1} - \frac{1}{2} \right) \tag{2.16.8}$$

and pair the operators in the manner preferred by the XY model. In zero external field, this is

$$H'_{H-F} = +J_z \sum_n (a_n^* a_{n+1} \langle a_{n+1}^* a_n \rangle + \text{H.C.}) - J_z \sum_n \left| \langle a_n^* a_{n+1} \rangle \right|^2 \tag{2.16.9}$$

with the change in apparent sign of J_z due to the anticommutation relations,
fermion statistics. (In an external field, pairings such as $\langle a_n^* a_n - 1/2 \rangle$
must also be retained.) It is easy to see that this problem involves some
self-consistency requirement, which are spelled out and solved in the recent
work of *Caliri* and *Mattis* [2.47a]. We show the specific heat calculated in
zero field for various J_z in Fig.2.19 taken from this work. The formulation
of the correct Hartree-Fock H' when the external field is present in (2.16.4)
is given in *Glauss* et al. [2.47b].

Amusingly, the spin 1/2 XY antiferromagnet (J < 0) can even be solved
exactly for in-plane orientations of the external field. This discovery by
Thomas and coworkers [2.48] is insufficiently understood. Briefly, starting
with (2.16.1) and J < 0 with **B** = (B,0,0) in the x-direction, they found $\langle S^x \rangle$
to be discontinuous in B at discrete values of B, with the last discontinuity
occurring at B = $\sqrt{2}$J, clearly a critical value of sorts. Although there is no

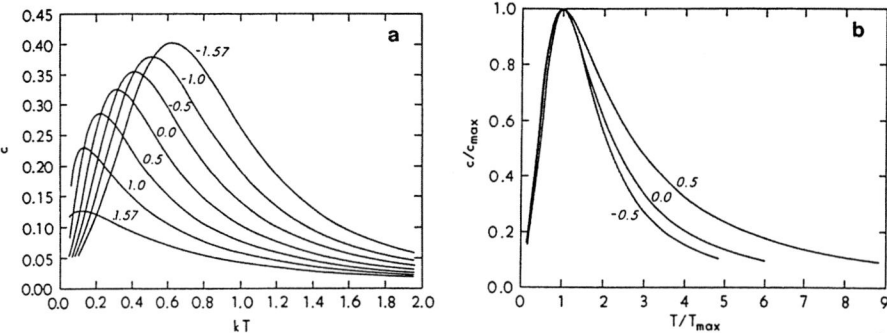

Fig.2.19. (a) Specific heat vs T for various ratios J_z/J_x in Hartree-Fock approximation to Heisenberg chain with $J_x = J_y \neq J_z$ [2.47a]. (b) Combining these results in terms of c/c_{max} vs T/T_{max} yields a universal curve only for $T/T_{max} \leqslant 1$

known systematic solution to the XY Hamiltonian when the field is not in the z-direction, these researchers *guessed* a product state of the form,

$$\psi = |\phi_1) \otimes |\phi_2) \otimes \cdots \qquad (2.16.10)$$

and found that it is the *ground state* of their H at $B = \sqrt{2}J$. In this expression, the individual spin states are

$$|\phi_n) = 2^{-\frac{1}{2}} \left\{ \exp[-(-1)^n(i\pi/8)]S_n^+ + \exp[(-1)^n(i\pi/8)] \right\} |\phi) \qquad (2.16.11)$$

taking $|\phi)$ to be the state of spin down. It is not known whether any of these ground state features persist to finite T, although it is an intriguing possibility.

The properties of the *classical XY model* are far easier to study than the extreme quantum s = 1/2 limit in the above. Again for the linear chain, we can obtain the thermodynamic functions exactly, using the method of *transfer matrices* which is particularly useful in 1D. For purposes of illustration, we specialize to the *planar model* (also called plane rotator model; the spins are constrained to the XY plane) and temporarily set B = 0. The effective Hamiltonian is then,

$$H = -J \sum_n \cos(\theta_n - \theta_{n+1}) \qquad (2.16.12)$$

where θ_n is the angle of the nth spin with respect to some arbitrary (fixed) axis in the XY plane.

As the total Hamiltonian consists of pieces $H_{n,n+1}$ which commute with one another, we write the partition function as an ordered sequence of integrations

$$Z = \mathrm{Tr}\left\{e^{-\beta H}\right\}$$

$$= \ldots \frac{1}{\pi} \int_{-\pi}^{\pi} d\theta_n \, \exp(-\beta H_{n,n+1}) \, \frac{1}{\pi} \int_{-\pi}^{\pi} d\theta_{n+1} \, \exp(-\beta H_{n+1,n+2}) \ldots \quad (2.16.13)$$

with the factors $1/\pi$ chosen so that Z is, arbitrarily, normalized to 2^N in the high-temperature limit. We view the evaluation of this repeated integration as an eigenvalue problem. For suppose we knew the eigenfunctions $f_j(\theta)$ and eigenvalues z_j in the integral equation,

$$\frac{1}{\pi} \int_{-\pi}^{\pi} d\theta' \, \exp[-\beta H(\theta,\theta')] f_j(\theta') = z_j f_j(\theta) \qquad (2.16.14)$$

then by iteration in Z, we obtain

$$Z = \sum_j (z_j)^N \qquad \text{(PBC)} \qquad (2.16.15a)$$

if the Nth spin is tied to the first (periodic boundary conditions). Or,

$$Z = 2z_0^{N-1} \qquad \text{(free ends)} \qquad (2.16.15b)$$

if the chain merely ends at the Nth spin (so there are only N-1 bonds). Here z_0 is the largest eigenvalue of (2.16.14), which is the only attainable one when we start with $f(\theta_1) = 1$.

These two results are actually the same in the thermodynamic limit, for the free energy with periodic boundary conditions is:

$$\ln Z_{PBC} = N \ln z_0 + \ln[1 + (z_1/z_0)^N + (z_2/z_0)^N + \ldots]$$

and may be compared to that with free ends:

$$\ln Z_{f.e.} = N \ln z_0 + \ln(2/z_0) \ .$$

The extensive part is the same in both cases, whereas the correction term, which is $O(1)$, expresses the thermodynamics of the particular end conditions.

With $H(\theta,\theta') = -J\cos(\theta - \theta')$, the integrand is well known as the generating function for Bessel functions I_n of (2.14.7):

$$\exp(\beta J \cos(\theta - \theta')) = \sum_{n=-\infty}^{+\infty} I_n(\beta J) \, \exp[in(\theta - \theta')] \ . \qquad (2.16.16)$$

It is of the form

$$V(\theta,\theta') = \sum_n \omega_n \psi_n^*(\theta') \psi_n(\theta) \ ,$$

i.e., a diagonal matrix in the represenation of its eigenfunctions $\exp(in\theta)$, with eigenvalues $\omega_n = I_n(\beta J)$. We call $V(\theta,\theta')$ the transfer matrix. The eigenvalues I_n are the Bessel functions of imaginary argument,

$$I_n(K) = \frac{1}{2\pi} \oint d\theta \, \exp(K\cos\theta + in\theta) = I_{-n}(K) \tag{2.16.17}$$

$$= i^n J_n(-iK)$$

satisfying usual Bessel function identities $I_1 = dI_0(K)/dK$, etc. Their magnitudes are in natural order: $I_0 > I_1 > I_2 > \ldots$ at all values of $K = \beta J$.

Consequently, the largest eigenvalue z_0 is always $2I_0(\beta J)$ and

$$Z = [2I_0(\beta J)]^N \quad , \quad F = -N \, kT \, \ln 2I_0(\beta J) \tag{2.16.18}$$

yields the thermodynamics in zero external field. In finite field, or in general if we change the form of $H_{n,n+1}$, we can expand the eigenfunctions of the new operators in terms of the complete set of ψ_n's, as in perturbation theory. Take the example of the external field,

$$H'_{n,n+1} = -\frac{1}{2} B(\cos\theta_n + \cos\theta_{n+1}) \quad . \tag{2.16.19}$$

The new transfer matrix can be written as the sum of three matrices: the unperturbed matrix, a matrix linear in B and one which is quadratic in B. For the calculation of the paramagnetic susceptibility all we require is the free energy to second order in B, thus we treat the linear perturbation in second-order perturbation theory, and the quadratic perturbation in first-order perturbation theory.

Explicitly,

$$\delta_1 V = 2 \exp[\beta J \cos(\theta - \theta')][(\tfrac{1}{2} \beta B)(\cos\theta + \cos\theta')] \tag{2.16.20}$$

for which second-order perturbation theory yields

$$\delta z = \sum_{n \neq 0} \frac{|(\delta_1 V)_{n,0}|^2}{2(I_0 - I_n)} \quad , \tag{2.16.21a}$$

while the term second-order in B is

$$\delta_2 V = 2 \exp[\beta J \cos(\theta - \theta')][\tfrac{1}{8} (\beta B)^2 (\cos\theta + \cos\theta')^2] \tag{2.16.22}$$

which, in first-order, yields

$$\delta z = (\delta_2 V)_{0,0} = \frac{1}{4} (\beta B)^2 (I_0 + I_1) \quad . \tag{2.16.21b}$$

The evaluation of (2.16.21a) yields $\frac{1}{4} (\beta B)^2 (I_0 + I_1)^2 / (I_0 - I_1)$, so combining it with (2.16.21b) we find for the new eigenvalue,

$$z = 2I_0 \left[1 + \frac{1}{4} (\beta B)^2 \frac{I_0 + I_1}{I_0 - I_1} \right] \quad . \tag{2.16.23}$$

We recognize $I_1/I_0 = d(\ln I_0)/dK$ as $\langle\cos(\theta_n - \theta_{n+1})\rangle \equiv \mu$, i.e., the nearest-neighbor correlation function in zero field ($K = \beta J$). Taking logarithms of z, and identifying the susceptibility through the expansion $f = f_0 - \frac{1}{2}\chi_0 B^2$, we arrive at the expression

$$\chi_0 = \frac{\mathbb{C}}{T}\left(\frac{I_0 + I_1}{I_0 - I_1}\right) = \frac{\mathbb{C}}{T}\left(\frac{1 + \mu}{1 - \mu}\right) \quad . \tag{2.16.24}$$

We should note that μ is an odd function of J, i.e., $\mu(-J) = -\mu(J)$. In the following chapter, we derive an identical expression for the linear-chain Ising model if we take $\mu = \tanh(\beta J)$, the appropriate correlation function in that case. For the isotropic classical Heisenberg model where $\mu = \coth(\beta J) - 1/(\beta J)$ (Problem 2.17), *Fisher* [2.49] also finds a susceptibility $\chi_0 = (\mathbb{C}/T)(1 + \mu)/(1 - \mu)$. These three examples suggest an interesting relationship between the susceptibility of a ferromagnet ($J > 0$) and antiferromagnet ($J < 0$) in 1D. If we define $X(\beta J) \equiv T\chi_0/\mathbb{C}$, they satisfy:

$$X(\beta J)X(-\beta J) = 1 \quad . \tag{2.16.25}$$

We do not know whether this symmetry is destroyed by quantum fluctuations, nor how generally it is satisfied in 1D, but (2.16.24) certainly expresses in a compact way the relation between long-range and short-range order in 1D.

The well-informed reader will already have noted that the sequence of integrations (convolution) (2.16.13) is a Feynman path-integral or functional integral, (with only the simple modification that n labels *physical* points and not an arbitrary sequence). In view of the relationship between such integrals and the Schrödinger equation, we should not be surprised at the relationship between the partition function and the transfer-matrix equation (2.16.14), one which has been particularly successful in the resolution of the two-dimensional Ising model treated in the next Chapter.

....

Problem 2.17. Identify the transfer matrix for the classical Heisenberg linear chain, and derive its expansion in terms of spherical harmonics eigenfunctions $Y_{\ell,m}(\theta,\phi)$ and eigenvalues $j_\ell(x) = (\pi/2x)^{\frac{1}{2}}J_{\ell+\frac{1}{2}}(x)$ (spherical Bessel functions):

$$V(\theta,\phi|\theta',\phi') = \sum_{\ell=0}^{\infty}\sum_{m=-\ell}^{+\ell} i^\ell j_\ell(i\beta J)Y^*_{\ell,m}(\theta,\phi)Y_{\ell,m}(\theta',\phi') \quad .$$

Calculate $\mu \equiv \langle S_n \cdot S_{n+1}\rangle = -\partial\ln z_0/\partial\beta J$ and compare with Fisher's result.

....

2.17 The XY Model: 2D

The XY model is an interesting example of cooperative phenomena in 2D systems with continuous symmetry: there is no LRO, yet a phase transition occurs at finite T_c.

The information on the extreme quantum limit $s = 1/2$ is sparse, much of it in numerical studies by *Betts* and his collaborators [2.50]. For indications concerning the low-temperature phase, they investigated the ground state susceptibility tensor χ_{ij}^0 (χ_{zz}^0 refers to magnetization in the z-direction owing to an applied field in the z-direction, χ_{xz}^0 to magnetization in the x-direction owing to an applied field in the z-direction, etc.), finding $\chi_{zz}^0 = 0$ while $\chi_{xx}^0 = \chi_{yy}^0$ both *diverge* for $J > 0$. For $J < 0$, the latter are finite.

The proof that the ground state χ_{zz}^0 vanishes is relatively simple, and can be extended by inspection to *all* components of the susceptibility tensor for the isotropic Heisenberg antiferromagnet. It is also valid in any dimension d. We write the ground-state susceptibility in terms of the matrix elements of the total spin operator $S_{tot}^z = \sum_i S_i^z$ connecting the ground state (0) to the excited spectrum (α):

$$\chi_{zz}^0 = \sum_{\alpha \neq 0} |(S_{tot}^z)_{\alpha,0}|^2 / (E_\alpha - E_0) \quad . \tag{2.17.1}$$

Because the ground state is an eigenfunction of S_{tot}^z with eigenvalue 0 (see [2.51] for a proof), this expression vanishes identically! While this does *not* prove that $\lim T \to 0 \; \chi_{zz}(T) = 0$, it *does* indicate the absence of long-ranged $\langle S_i^z S_j^z \rangle$ correlations in the ground state.

We briefly touched upon the need to consider *vortices* in the XY model in [Ref.2.1, Sect.5.15]. A plausible vortex *operator* for 4 spins at the corners of an elementary square is [2.50]:

$$V_Q = \frac{1}{4} (S_1^x S_2^y - S_2^y S_3^x + S_3^x S_4^y - S_4^y S_1^x) \quad . \tag{2.17.2}$$

Of the 16 eigenstates of these 4 spins, V_Q has zero eigenvalue on 12, +1 on two and -1 on the remaining two. The expectation value of V_Q is zero in general, by symmetry, but its square V_Q^2 indicates the vortex-antivortex density. Numerical studies show this to be relatively small and increasing slowly with temperature, until a finite T_c is reached. Thereafter, their number increases rapidly to the theoretical maximum [2.50].

A rather more round about approach to the $s = 1/2$ XY model in 2D has been taken by *Lagendijk* and *De Raedt* [2.52], who by use of Trotter's formula

$$e^{A+B} = \lim_{m \to \infty} (e^{A/m} e^{B/m})^m$$

disentangle all the even-numbered sites from the odd-numbered sites in the evaluation of the partition function, which they perform in the $m=1$ approximation to the above limit. Their calculation is nontrivial, as it requires a mapping onto the 8-vertex model, for which solutions are known only in certain cases. They conclude that the $m=1$, 2D XY model has a logarithmically divergent specific heat at T_c, with T_c given by $\sinh(2\beta_c J) = 1$, i.e., $kT_c = (2.27...)J$, just as in the corresponding Ising model. Obviously, there is need for additional work in this area. The $s=1/2$ XY model is the prototype of the hard sphere Bose gas. As such, our increased understanding of its thermodynamic properties in 2D and 3D will be directly relevant to continued progress in the related theory of superfluidity.

But undoubtedly it is the *planar* (*alias* plane rotator) model,

$$H = -J \sum_{(i,j)} \cos(\phi_i - \phi_j) \quad , \tag{2.17.3}$$

which has generated the greatest activity following *Kosterlitz* and *Thouless'* [2.53] discovery of its remarkable properties. As it is not yet clear which of several approach will ultimately prove the most fruitful, we indicate several paths to the resolution of this model. In the original semiphenomenological approach, bolstered by corrections due to scaling and renormalization group arguments [2.54], one separates the vortex (topological) excitation degrees of freedom from the magnon (small deviations about topological configurations) normal modes. It is the former which are responsible for the anomalies, including the phase transition, while the latter provide an analytic, essentially uninteresting background (as is evident if we recall the two-dimensional spherical model, Sect.2.9).

Then, instead of the total H above, one considers primarily the vortex Hamiltonian consisting of two parts: the energy of creating each vortex, and the interaction energy. At low temperature, pairs of vortices of opposite sign are created at small distance from each other at a small cost in energy. At higher temperatures, they interpenetrate and form a fluid formally analogous to an electrolyte. Finally, above T_c, free vortices are created in ever increasing numbers with increasing temperature. The energy to create a single vortex was estimated in [Ref.2.1, Sect.5.15] (or see [2.53,54]):

$$\Delta E = 2\pi J \ln(R/a) \tag{2.17.4}$$

where R is the radial dimension of the sample and a the lattice parameter. The entropy gained in this process is

$$\Delta \mathscr{S} = k \ln(R/a)^2 \tag{2.17.5}$$

and thus the excess free energy is,

$$\Delta F \approx 2(\pi J - kT) \ln(R/a) \quad . \tag{2.17.6}$$

The phase transition occurs when vortices are spontaneously generated, i.e., at approximately

$$kT_c \approx \pi J \quad . \tag{2.17.7}$$

The important corrections come from vortex interactions:

$$\Delta E_{ij} = -2\pi J q_i q_j \ln R_{ij}/a \tag{2.17.8}$$

and interactions with background magnon terms. Their result is to change J in (2.17.1) to $\frac{1}{2} J_{eff}(T)$ [2.54].

The straightforward evaluation of this partition function is more diffi-cult than one would believe. A significant attempt was made by *Villain* [2.55], whose modifications are referred to as the Villain model. Recognizing that the effective interactions will be logarithmic (so that the Boltzmann factors will become temperature-dependent *powers* of the vortex separations) he starts with an analysis of the Boltzmann factors through formal expansion. Consider the expansion:

$$\exp(2\beta J \cos\phi) \doteq \text{const.X} \sum_{n=-\infty}^{+\infty} \exp[-\beta A(\phi - 2\pi n)^2] \tag{2.17.9}$$

with the factor 2 to allow for double summation over pairs in subsequent cal-culations. The parameter A is a function of T, which together with the con-stant multiplying the sum above, is adjusted for a best fit. Both sides are formally periodic in ϕ, but in evaluating the partition function ϕ is now taken over the entire range - , + instead of the fundamental interval. The justification is that for any periodic function $F(\phi)$,

$$\int_{-\pi}^{+\pi} d\phi F(\phi) = \lim_{\varepsilon \to 0} 2(\beta\pi\varepsilon)^{\frac{1}{2}} \int_{-\infty}^{+\infty} d\phi \, \exp(-\beta\varepsilon\phi^2)F(\phi) \quad .$$

The simplification is that the partition function will contain only Gaus-sian-type integrations, calculable in closed form. The effective Hamiltonian for Villain's model is thus,

$$H_{eff} = \frac{1}{2} \sum_{i,j} A(\phi_i - \phi_j - 2\pi n_{ij})^2 + B \sum_{i} (\phi_i - 2\pi \nu_i)^2 \tag{2.17.10}$$

in which the $n_{ij} = -n_{ji}$ and the ν_i are discrete field variables, which supplement the ϕ_i. Pursuing this, Villain obtained the properties of the planar model with a critical temperature half that in (2.17.7), more in line with series estimates.

This approach was taken an important step further by José et al. [2.56], who found a neat trick for performing sums over integers n_{ij} in the partition function, and were then enabled to derive and rederive all known results by reasonably straightforward calculations. A slight modification of Villain's expansion results in complex exponentials, and reduces the evaluation of Z in zero external field to

$$Z = \int_{-\infty}^{+\infty} \prod_i d\phi_i \prod_{ij} \sum_{n_{ij}=-\infty}^{+\infty} \exp\left\{ \sum_{i,j} [F(n_{ij}) + in_{ij}(\phi_i - \phi_j)] \right\} . \qquad (2.17.11)$$

The ϕ_i's are now eliminated. One obtains a partition function purely in terms of integer quantum numbers, which we can then identify with the vortices. Each integration over a ϕ_i forces the integers with which it interacts to add to zero, a form of momentum conservation. Let us examine this with the aid of Fig.2.20, in which we label the neighbors of a given site i by 1,2,3,4.

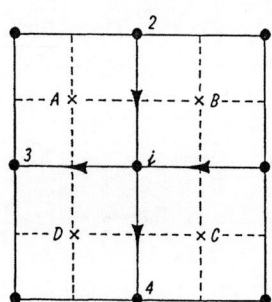

Fig.2.20. Construction of a dual lattice (A,B,C,...) to satisfy curl-free condition on lattice (i,1,2,...)

Integrating over ϕ_i alone, we obtain zero unless $n_{1i} + n_{2i} - n_{i3} - n_{i4} = 0$. We represent these integers by arrows, and vertify that 2 arrows go in, and two go out of the vertex at i. This discrete version of a zero divergence condition can be satisfied by the choice of a dual field of integers n_j, located within each plaquette of the original lattice as shown in the Figure. Then setting

$$n_{i1} = n_B - n_C , \quad n_{i2} = n_A - n_B ,$$

$$n_{i3} = n_A - n_D \text{ and } n_{i4} = n_C - n_D , \qquad (2.17.12)$$

around i and making similar identifications throughout, the desired conditions are met identically for every point in the original lattice, and one obtains a partition function over the integers of the dual lattice. Finally, one makes use of the Poisson summation formula for each integer n_A:

$$\sum_{n=-\infty}^{+\infty} g(n) = \sum_{m=-\infty}^{+\infty} \int_{-\infty}^{\infty} d\phi \, g(\phi) \, \exp(-2\pi i m \phi) \qquad (2.17.13)$$

transforming Z into the form:

$$Z = \prod_i \sum_{m_i=-\infty}^{+\infty} \int_{-\infty}^{+\infty} d\phi_i \left\{ \exp\left[\sum_{i,j} F(\phi_i - \phi_j) + \sum_j 2\pi i m_j \phi_j \right] \right\} \qquad (2.17.14)$$

with the sums over the dual lattice points. In this representation, the ϕ_j's describe the magnon modes, and the m_j's the vortex quantum numbers. Especially at low temperatures, the sums converge quickly and the integrals (F is not a very complicated function) can be performed by a variety of methods. The results of Villain are obtained, but with a renormalized J_{eff}. We summarize them as follows.

The spin-spin correlation function decreases non-exponentially, as a power law

$$\langle\cos(\theta_0 - \theta_r)\rangle = \left(\frac{a}{r}\right)^{\eta(T)} \quad \text{with} \qquad (2.17.15)$$

$$0 \leqslant \eta(T) = kT/2\pi J_{eff} \leqslant \frac{1}{4} \, , \qquad (2.17.16)$$

the phase transition occurring at $kT_c = \frac{1}{2} \pi J_{eff}(T_c)$. Above T_c the decay is exponential with a finite correlation length $\xi(T) \approx C \exp[b/(T - T_c)^\nu]$ (with $\nu \approx 0.7$ [2.57] or $\nu = 0.5$ [2.54]), C and b being constants; the magnetic susceptibility χ is proportional to ξ at $T \gtrsim T_c$, and both are essentially infinite for $T \leqslant T_c$ [2.57].

This behavior differs considerably from models with LRO below T_c. Because exponents are continuously varying up to T_c, one describes the phase transition as a line of critical points from 0 to T_c, rather than as a unique phase transition at T_c. The main surprise is that $\eta_c = 1/4$, far from the theoretical maximum $\eta = 2$, discussed in (2.15.1), but the same as for the Ising model in 2D. The power-law behavior is precisely what one expects from free magnons with dispersion $\omega(q) = \frac{1}{2} J_{eff} q^2$, as we now show.

$$\langle\cos(\theta_0 - \theta_r)\rangle = 1 - \frac{1}{2} \langle(\theta_0 - \theta_r)^2\rangle + \ldots$$

$$\approx \exp\left[-\frac{1}{2} <(\theta_0 - \theta_r)^2>\right] = \exp[-G(r)] \qquad (2.17.17)$$

with

$$G(r) = \frac{kT}{J_{eff}N} \sum_q (1 - e^{iq\cdot r})q^{-2} \approx \frac{kT}{2\pi J_{eff}} \ln(r/a) \qquad (2.17.18)$$

from which (2.17.15) follows.

These properties and the specific heat were examined numerically by Monte Carlo simulation (which generally confirms the theoretical picture). We reproduce the numerical experiments on the specific heat by *Tobochnik* and *Chester* [2.57] in Fig.2.21. The interesting features are a specific heat maximum at $kT = 1.02$ J, with the phase transition occurring at the lower teperature $kT_c = 0.89$ J. There appear to be no specific heat anomalies at T_c (defined as the temperature at which the susceptibility diverges).

Is it possible that, finally, a simple theory can yield this plethora of interesting results? In the following section we prepare for this contingency, by deriving the transfer matrix in a rather transparent form [2.58].

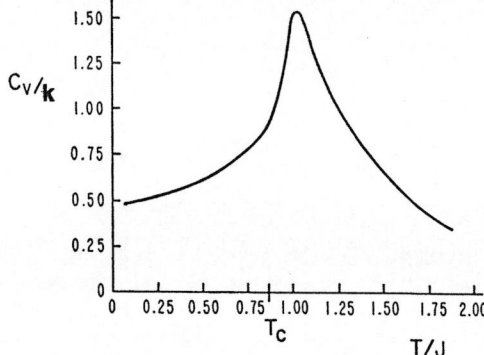

Fig.2.21. Specific heat in planar rotator model in 2D, as obtained by numerical (Monte Carlo) experiments. (——) replaces experimental points published by Tobochnick and Chester [2.57] to better show some features: maximum near $kT = 1.02$ J, smooth phase transition at $kT_c = 0.89$ J

2.18 Transfer Matrix of Plane Rotator Model

The evaluation of the partition function of the 2D or 3D plane rotator model is reducible to an eigenvalue problem, that of obtaining the largest eigenvalue of the transfer matrix. We derive this matrix in simplest possible form [2.58], but its complete resolution is another matter! [2.59].

Let us illustrate with two connected chains, to see what needs be generalized in the earlier formulation of the one-dimensional problem (Sect.2.16). Label the coordinates on the first chain θ_n and on the second, ϕ_n. Then,

$$H = \sum_n^N H_{n,n+1} \quad \text{where} \tag{2.18.1}$$

$$H_{n,n+1} = -J_1[\cos(\theta_n - \theta_{n+1}) + \cos(\phi_n - \phi_{n+1})] - J_2 \cos(\theta_{n+1} - \phi_{n+1}) \tag{2.18.2}$$

in which we distinguish horizontal (intrachain) bonds J_1 from vertical (interchain) bonds J_2 for greater generality. The coupling constants J_1 and J_2 can be set equal at the end of the calculation, if desired.

Writing Z as an ordered sequence of integrations —similar to (2.16.13) — we find it is given by $Z = z_0^N$, with z_0 the maximal eigenvalue of the integral equation:

$$\frac{1}{\pi^2} \int\limits_{-\pi}^{+\pi}\!\!\int d\theta_n d\phi_n \, \exp(-\beta H_{n,n+1}) \chi(\theta_n, \phi_n) = z_0 \chi(\theta_{n+1}, \phi_{n+1}) \quad . \tag{2.18.3}$$

The transfer matrix $V_{n,n+1} = 4 \exp(-\beta H_{n,n+1})$ is of the form $V^{(1)} V^{(2)}$, where

$$V^{(1)}_{n,n+1} = 4 I_0^2(\beta J_1) \sum_{p,q=-\infty}^{+\infty} \exp\{-\beta[w(p) + w(q)]\} \exp[ip(\theta_n - \theta_{n+1})]$$

$$\times \exp[iq(\phi_n - \phi_{n+1})] \tag{2.18.4}$$

represents the two individual chains, according to (2.16.16), with

$$w(p) \equiv kT \ln[I_0(\beta J_1)/I_p(\beta J_1)] \tag{2.18.5}$$

defined for convenience, and where $V^{(2)}$ contains the interchain links:

$$V^{(2)}_{n,n+1} = \exp[\beta J_2 \cos(\theta_{n+1} - \phi_{n+1})] \quad . \tag{2.18.6}$$

Performing the integration indicated in (2.18.3) we obtain the eigenvalue equation in the form of an exponentiated differential equation:

$$4 I_0^2(\beta J_1) \exp\{-\beta[w(p_\theta) + w(p_\phi)]\} \exp[\beta J_2 \cos(\theta - \phi)] \chi = z_0 \chi \tag{2.18.7}$$

with $p_\theta = -i\partial/\partial\theta$ and $p_\phi = -i\partial/\partial\phi$. This can be simplified further, writing $z_0 = 4 I_0^2(\beta J_1) \exp(-\beta f)$, with f (the "interaction free energy") the new eigenvalue. As usual, we want only the *lowest* f. The equation now reads:

$$\exp\{-\beta[w(p_\theta) + w(p_\phi)]\} \exp[\beta J_2 \cos(\theta - \phi)] \chi(\theta,\phi) = \exp(-\beta f) \chi(\theta,\phi) \quad . \tag{2.18.8}$$

....
Note. To fix ideas, we discuss the behavior of $w(p)$. From series and asymptotic expansions we can analyze $w(p)$ at small $\beta J_1 = K_1$ and also at large K_1. In the vicinity of the 2D phase transition, K tends to be O(1) however, so there the exact definition of the I_p must be used in (2.18.5).

High temperature $(K_1 \to 0)$: $I_p(K_1) \approx (\frac{1}{2} K_1)^p / \Gamma(p + 1)$, thus

$$w(p) = kT|p|[\ln(2kT|p|/eJ_1)] \quad . \tag{2.18.5a}$$

This formula is also applicable at large $|p| \gg 1$, for arbitrary K_1.

Low temperature $(K_1 \to \infty)$: the asymptotic expansions [2.60] again lead to (2.18.5a) for large $|p|$, but at small $|p|$ we find:

$$w(p) = \frac{1}{2} (kT)^2 p^2 / J_1 \quad . \tag{2.18.5b}$$

In general, $w(p)$ is a smoothly increasing function of $|p|$, with $w(0) = 0$.
....

We would like now to solve the eigenvalue problem. First, for some insight into the nature of the solutions, we combine the exponents of $v^{(1)}$ and $v^{(2)}$ and simply solve:

$$[w(p_\theta) + w(p_\phi) - J_2 \cos(\theta - \phi)]\chi = f\chi \tag{2.18.9}$$

for its ground-state eigenvalue. There is no proof that this leads to an accurate solution, but the equivalent approximation in the Ising model transfer matrix (Chap.3) yields excellent results. Because the interaction conserves angular momentum, we may write $\chi(\theta, \phi)$ in the form:

$$\chi(\theta, \phi) = \exp[iQ(\theta + \phi)] \sum_k A_k \exp[ik(\theta - \phi)] \tag{2.18.10}$$

and set Q = 0 in the ground state. The eigenvalue equation now takes the form of a difference equation

$$[2w(k) - J_2 - f]A_k = \frac{1}{2} J_2(A_{k+1} + A_{k-1} - 2A_k) \quad , \tag{2.18.11}$$

a reasonably simple equation to analyze (it is the equation of a particle hopping on a linear chain labeled by integers k, subjected to a potential $2w(|k|) - J_2$ which has only bound states) if not to solve explicitly. Certainly, f and its derivatives are continuous in the parameters J_1, J_2 and T, and there can be no phase transition. This is expected as there are only *two* chains.

The solution by means of (2.18.10) of the *exact* equation (2.18.8) takes on a more formidable mien. It is convenient to left-multiply both sides of the equation by $\exp \beta[w(p_\theta) + w(p_\phi)]$, and express $\exp(\beta J_2)$ in a series expansion involving the Bessel functions. We obtain

$$I_0(\beta J_2) \sum_{q=-\infty}^{+\infty} \exp[-\beta \hat{w}(q)]A_{k-q} = \exp\{-\beta[f - 2\hat{w}(k)]\}A_k \qquad (2.18.12)$$

with

$$\hat{w}(q) = \hat{w}(-q) = kT \ln[I_0(\beta J_2)/I_q(\beta J_2)] \qquad (2.18.13)$$

by analogy with (2.18.5). This equation is of a less familiar type, but analysis of it indicates that the solution is even smoother function of the
parameters, and that f agrees with the solution of the approximate equation
(2.18.11) both at low and at high temperatures.

Without entering the domain of numerical analysis of the above, we now
turn to the *full* 2D transfer matrix. Its formulation follows by simple extension of the preceding. We define the nth line (2D) [or n^{th} plane (3D)],
to contain all the spins at the nth position of the linear chains. They are
labeled θ_i (in the example of the two chains, $\theta_1 = \theta$ and $\theta_2 = \phi$) with
$i = 1, 2, \ldots, N'$. θ_i interacts only with $\theta_{i\pm 1}$, its nearest-neighbors on this
line. (In 3D, *two* subscripts label a spin in the plane of the transfer matrix,
and again, only nearest-neighbors interact.) As $N' \to \infty$, a phase transition is
allowed. Now, we factor $2^{N'} I_0^{N'}(\beta J_1)$ from the largest eigenvalue,

$$Z = z_0^N, \quad \text{and} \quad z_0 = 2^{N'} I_0^{N'}(\beta J_1) \exp(-\beta f N') \qquad (2.18.14)$$

in which form f is the interaction free energy *per* spin. The eigenvalue
equation is now:

$$V\chi(\theta_1, \theta_2, \ldots \theta_{N'}) = \exp(-\beta f N')\chi(\theta_1, \theta_2, \ldots \theta_{N'}) \qquad (2.18.15)$$

where $V = V^{(1)}V^{(2)}$, and

$$V^{(1)} = \exp\left[-\beta \sum_{n=1}^{N'} w\left(\frac{1}{i} \partial/\partial\theta_n\right)\right] \equiv \exp[-\beta H^{(1)}] \qquad (2.18.16a)$$

and

$$V^{(2)} = \exp\left[\beta J_2 \sum_{n=1}^{N'} \cos(\theta_n - \theta_{n+1})\right] \equiv \exp[-\beta H^{(2)}] \qquad (2.18.16b)$$

in 2D. [In 3D the sum in (a) ranges over the 2D plane of the transfer matrix,
and in (b) over the nearest-neighbor pairs in that plane.] The solution of
(2.18.15) yields the *exact* thermodynamic properties of the model. Once again,
it is instructive to examine the simpler model eigenvalue problem:

$$[H^{(1)} + H^{(2)}]\chi = (f N')\chi \qquad (2.18.17)$$

which is recognizable as a linear chain of coupled pendulums, subject to zero-
point motion induced by the kinetic energy $w(p_i)$. The kinetic energy is the

expression in the transfer matrix of the temperature T in the original 2-dimensional problem; the higher the physical parameter T, the higher the kinetic energy parameter in the present equation. Thus, as temperature is raised, the equivalent of the uncertainty principle causes the pendulums to become misaligned. At high T each pendulum is approximately homogeneously distributed over the interval $0 - 2\pi$, and is essentially uncorrelated to the position of its neighbors. At T_c the correlation to nearest-neighbor begins, until at $T = 0$ all the pendulums are at the same angle $\theta_1 = \theta_2 = \dots = \theta_{N'}$, which we can take to be $\theta = 0$. The value of T_c found by solving (2.18.17) [2.59] satisfies:

$$(2kT_c/J)\ln(2kT_c/J) = 1 \qquad\qquad\qquad (2.18.18)$$

which yields: $kT_c = 0.88$ J, in good agreement with the numerical "experiments" [2.57] for 2D.

The statement of the problem is essentially the same in 3D. The transfer matrix is in 2D, the pendulums interact with a greater number of neighbors and one expects a greater degree of order at low T. The examination of the approximate equation (2.18.17) or of the exact equation (2.18.15) is in either case (2D or 3D) beyond the scope of this book, although the reader who has persevered this far will possess the necessary technical tools. The solution of the *two-dimensional* plane-rotator model, obtained by the method in this chapter, involves a number of manipulations published separately in [2.59], where the interested reader may find the necessary details. In the following chapter, we turn to a similar problem in connection with the two-dimensional Ising model. There, the transfer matrix is easily expressible in terms of free fermions, and the exact solution of that model *will* be explicitly demonstrated.

3. The Ising Model

In this prototype theory of ferromagnetism —and of many other physical phenomena as well —a spin $S_i = \pm 1$ is assigned to each of N sites on a fixed lattice. The spins, which live on the vertices of the lattice, interact with one another by means of bonds (the links of the lattice). These have strengths J_{ij} in energy units. In addition, the spins can interact with external fields B_i of arbitrary strengths. The total energy is then given by:

$$H = - \sum J_{ij} S_i S_j - \sum B_i S_i$$

and can be directly evaluated in any of the 2^N spin configurations. In the most familiar version of the Ising model, the interactions are limited to nearest-neighbors on the lattice and the magnetic field is homogeneous, $B_i = \text{constant}$.

An historical review of this model is of some interest, as it illustrates the frequently devious paths along which research proceeds in such matters. If the reader's appetite for history is whetted, he will find a rather more thorough account in [3.1].

While the idea of a microscopic theory in which elementary spin dipoles are constrained to quantized directions dates back to *Lenz* in 1920 [3.2], the calculational burden fell upon his student Ising some years later. But *Ising*'s Ph.D. thesis and later publication [3.3] was not very encouraging. Based as it was on a one-dimensional analysis, it revealed no ferromagnetic phase above the absolute zero of temperature. The ground state was ordered: all spins parallel (all up or all down), yet the long-range order disappeared at any finite positive temperature. Thus, despite the nearest-neighbor bonds, this model displayed only paramagnetic behavior at finite T. Concerning his thesis, Ising has written:

At the time...Stern and Gerlach were working in the same institute (Hamburg) on their famous experiment on space quantization...I discussed the results of my paper widely with Prof. Lenz and Dr. Wolfgang Pauli, who at that time was teaching in Hamburg. There was some disappointment that the linear model did not show the expected ferromagnetic properties [3.1].

As a consequence, interest in Lenz' model died out, while other avenues of inquiry were opened, such as Heisenberg's fully quantum-mechanical model and the itinerant electron theories of Bloch and, later, Stoner. For it was commonly held that Lenz' semiclassical model was incapable of explaining the known laws of ferromagnetism, including the phase transition at the Curie temperature.

Ultimately, the study of what came to be known as Ising's model was revived by a parallel development in alloy physics. The Bragg-Williams molecular-field theory of the thermodynamics of alloys, patterned after Weiss, was justified by cluster calculations by *Bethe* [3.4] on a model which was mathematically, if not physically, identical to that of Lenz: If in the theory of an AB alloy, one allows spin up to refer to an atom of type A and spin down to an atom B, arranging the magnitudes of the bonds J_{ij} and of the external field B so as to correspond to realistic AA, AB and BB bonding energies and chemical potentials, he found a perfect correspondence to the Ising model in an external field. The solution of one entails the other. This approach to the structure of alloys is incorporated into metallurgy theory, appearing already in the 1939 text *Statistical Thermodynamics* of *Fowler Guggenheimer* [3.5]. In an alternative application, the B's can represent vacancies in a solid of A's. When sufficiently dilute, the latter form a lattice gas, which may condense at sufficiently high density or low temperature. Thus, the study of the Ising model contains the seeds of a first-principles theory of the phases of nonmagnetic matter: solid, liquid and gas.

The phase diagram of the lattice gas has been obtained by *Lee* and *Yang* in two remarkable papers [3.6], based on the large amount of information known about the Ising model. The equation of state they obtained gave the mixed phase region correctly, without invoking Maxwell's construction —which had previously been thought to be unavoidable. Lee and Yang also drew attention to the importance of extending the model parameters, such as the external field, into the complex plane. Their study of the zeros of the partition function in complex fields created a new tool for the study of phase transitions.

However, Lee and Yang were not first to treat the Ising model of magnetism (and the lattice-gas problem) within a unified formalism. Right after *Bethe*'s work [3.4], *Peierls* recognized that ferromagnetism, order-disorder transformations in alloys, and the very existence of phase transitions in microscopic theory could all be reduced to the examination of the Ising model in two, or three dimensions —but *not* in one [3.7]! Peierls contributed an important theorem that in two dimensions or higher, long-ranged order (LRO) persists at sufficiently low temperature. Conversely, LRO is known to disappear above some critical temperature, when there is sufficient thermal disorder. Thus, Peierls established the relevance of the Ising model to the Curie problem. In the technicalities of his proof of this theorem, Peierls erred and was ultimately corrected by *Griffiths* [3.8] some 28 years later.

The exact solution of the thermodynamic properties of the two-dimensional Ising model was the consequence of innovative research by several individuals, working out ideas which culminated in the exact calculation by *Onsager* [3.9] of the partition function of this model (in the absence of an external field). In 1941, *Kramers* and *Wannier* [3.10], assuming the existence of a unique phase transition, located T_c *exactly* by means of a duality transformation they discovered. Such a transformation maps the high- and low-temperature properties of the partition function into one another; the fixed point of this mapping is then T_c. These authors, invoking what is now termed "finite-size scaling" arguments, derived a logarithmically divergent specific heat anomaly at T_c, quite unlike anything which has been obtained heretofore in statistical mechanics. But it was *Onsager* [3.9] who, in 1944, provided the exact calculations. The tale continues in [Ref.3.1] and in the following pages. We terminate these prefactory remarks by noting that the prospects of solving the two-dimensional Ising model in an external magnetic field, or of solving the three-dimensional Ising model, remain as tantalizingly elusive today as they were in 1925, despite an impressive array of methods and theorems dedicated to these tasks: a clear challenge to the present reader!

3.1 High Temperature Expansions

In general, high-temperature series can be generated for *any* spin Hamiltonian by straightfoward expansion of the exponential in a Taylor series:

$$Z_N(\beta) = Z_N(0)<1 - \beta H + \frac{1}{2} \beta^2 H^2 - \ldots>_0 \qquad (3.1.1)$$

where $<>_0$ stands for an unweighted average over all spin parameters, or equivalently, for a thermal average at $\beta = 0$. But regardless of how small β may become (or, however, high the temperature) each succeeding term is one order of N greater than the preceding so that this series clearly diverges in the thermodynamic limit! This difficulty is resolved by noting that Z_N must be of the form $\exp(-\beta F)$, where $F = fN$ and f is independent of N. Thus the right-hand side must *also* be precisely exponential in N. (Contributions which do not lead to an extensive F but are, say, higher-order in N would destroy the existence of a thermodynamic limit, while terms lower-order in N would not survive the thermodynamic limit).

Let us test this by expanding $\log Z_N$, suitably regrouped into contributions for each power of β [NB: $\log \equiv \ln$ is *always* base e here]:

$$\log Z_N(\beta) = \log Z_N(0) - \beta[<H>_0] + \frac{1}{2}\beta^2\left[<H^2>_0 - <H>_0^2\right]$$

$$- \frac{1}{3!}\beta^3\left[<H^3>_0 - 3<H^2>_0<H>_0 + 2<H>_0^3\right]$$

$$+ \frac{1}{4!}\beta^4\left[<H^4>_0 - 4<H^3>_0<H>_0 - 3<H^2>_0^2 + 12<H^2>_0<H>_0^2 - 6<H>_0^4\right]$$

$$- \dots + \frac{1}{m!}(-\beta)^m[\ ]_m + \dots \tag{3.1.2}$$

The brackets $[\ ]_m$ are called m^{th}-*order cumulants*. For finite-ranged interactions J_{ij}, each cumulant may be verified to be $O(N)$, so that the summation yields a free energy which is extensive, term by term. Through comparison of the above series, one can develop a formula for the cumulants:

$$[\ ]_m = m! \sum_{\{n\}_m} (-1)^{M-1}(M-1)! \prod_{i=1}^{m} \frac{1}{n_i!}\left(\frac{1}{i!}<H^i>_0\right)^{n_i} \tag{3.1.3}$$

in which $\{n\}_m$ stands for a distinct set of non-negative integers n_i satisfying the condition $\sum_i i n_i = m$. The sum in (3.1.3) is over all such sets; within each set, $M \equiv \sum_i n_i$.

....

Problem 3.1. Use the above formula to generate the given cumulants, (3.1.2), and a new one: $[\ ]_5$.

....

Exercise: As a simple application of the above, assume each $S_i = \pm 1$ (spins one-half) and let the magnetic field be B, a constant. We calculate F to $O(T^{-2})$ for arbitrary J_{ij}. Because $<S_i>_0 = 0$, terms involving single spins vanish in all orders, whereas $S_i^2 = 1$. With these simplifications, we obtain:

$$F = -kTN\ln 2 - \frac{1}{2}\beta\left(\frac{1}{2}\sum_{ij}J_{ij}^2 + NB^2\right) - \frac{1}{6}\beta^2\left(\sum_{ijk}J_{ij}J_{jk}J_{ki} + 6B^2\sum_{ij}J_{ij}\right) + \dots$$

$$\tag{3.1.4}$$

The calculation of the induced magnetization yields some plausible results. Differentiating the above series term by term, with the Boltzmann constant k and the Curie constant $\mathbb{C}_{\frac{1}{2}}$

$$\mathcal{M} = -k\mathbb{C}_{\frac{1}{2}}\partial F/\partial B = N\mathbb{C}_{\frac{1}{2}}B\frac{1}{T}\left(1 + \frac{2}{kT}\frac{1}{N}\sum_{ij} J_{ij} + \ldots\right)$$

which, to the given order, is the same as

$$= N\mathbb{C}_{\frac{1}{2}}B\frac{1}{T}\left(1 - \frac{2}{kT}\frac{1}{N}\sum_{ij} J_{ij} + \ldots\right)^{-1}$$

$$= \frac{N\mathbb{C}_{\frac{1}{2}}B}{T - \theta} \quad . \tag{3.1.5}$$

Here, $\theta = k^{-1}\frac{2}{N}\sum J_{ij}$ is the same as T_G, (2.10.3), or T_c in the MFT if the bonds J_{ij} are ferromagnetic (positive) on the average, and is $-T_N$, the negative of the Néel temperature, if they are antiferromagnetic on average. It is amusing to recover this simple formula from the leading terms in the high-temperature expansion. Unfortunately, it is the asymptotic terms in the expansion which determine the behavior near T_c or θ, so the above is valid only for $T \gg \theta$.

. . . .

Problem 3.2. Decomposing an arbitrary Hamiltonian into two parts: $H = H_0 + \delta H$, where H_0 and δH commute, define $Z_0 = \text{Tr}\{\exp(-\beta H_0)\}$ and $<\delta H>_0 \equiv Z_0^{-1} \text{Tr}\{\delta H \exp(-\beta H_0)\}$.

Show that the total free energy $F = -kT \log \text{Tr}\{\exp(-\beta H)\}$ is given by a series similar to (3.1.2):

$$F = -kT \log Z_0 + <\delta H>_0 - \frac{kT}{2!}\beta^2\left[<(\delta H)^2>_0 - <\delta H>_0^2\right] + \ldots - \frac{kT}{m!}(-\beta)^m[\ ]_m \ldots$$

where $[\ ]_m$ is given by (3.1.3) with H^i replaced by $(\delta H)^i$. This perturbative expansion may be useful in cases where δH is complicated but small by comparison with H_0, provided Z_0 and $<>_0$ are readily calculable.

Problem 3.3. Use the above to derive the following identity for the zero-field susceptibility of an arbitrary system in an external field $B \to 0$:

$$\chi_0 = \frac{1}{kT}\left(<(\partial H/\partial B)^2>_0 - <\partial H/\partial B>_0^2\right) \quad .$$

Compare with (2.2.15).

. . . .

Although the equations above apply to arbitrary Hamiltonians and arbitrary spin magnitudes, there is considerable advantage and simplification in specializing to an Ising model, spins $\frac{1}{2}$ ($S_i = \pm 1$; thus $S_i^2 = 1$, $S_i^3 = S_i$, etc.). The exponentials in Z_N then can be easily resummed:

$$\exp(\beta J_{ij} S_i S_j) = \cosh \beta J_{ij} + S_i S_j \sinh \beta J_{ij}$$

and

$$\exp(\beta B_i S_i) = \cosh \beta B_i + S_i \sinh \beta B_i \quad . \tag{3.1.6}$$

Z_N is seen to be a polynomial in quantities denoted w:

$$w_{ij} = \tanh \beta J_{ij} \quad \text{and} \quad w_i = \tanh \beta B_i \quad , \tag{3.1.7}$$

after factorization of the related $\cosh \beta J_{ij}$ and $\cosh \beta B_i$ which, by themselves, are the leading contribution to the high-temperature partition function. Explicitly:

$$Z_N = \left(\prod_{ij} \cosh \beta J_{ij} \right) \left(\prod_i \cosh \beta B_i \right) \mathrm{Tr} \left\{ \prod_{ij} (1 + S_i S_j w_{ij}) \prod_n (1 + S_n w_n) \right\} \quad . \tag{3.1.8}$$

The trace yields a power series in the w's:

$$\mathrm{Tr}\{ \ \} = 2^N \left(1 + \sum_{ij} w_{ij} w_i w_j + \ldots \right) \tag{3.1.9}$$

which converges better than the original series (3.1.1). Even when βJ_{ij} or βB_i are large the w's can never exceed magnitude unity. We note that at finite N, the series (3.1.9) has a finite number of terms, unlike (3.1.1) which is always an infinite series. Being a polynomial in the w's, Z_N is *analytic* in β, hence cannot exhibit a phase transition. To study the expected phase transition at or near T_c, it will be necessary to *first* take the limit $N \to \infty$, then to calculate the contributions of the asymptotic (arbitrarily high) powers of the w's.

The evaluation of Z_N appears at first sight to be as complicated in 1D as it is in 2D or 3D. However, a plethora of tricks enable us to solve the 1D problem trivially. Another variety of methods can be brought to bear on the substantially more difficult 2D problem, but solve it only in vanishingly small external field. For the 2D model in external field, or in 3D, our only known recourse is to numerical analysis and approximate methods. For these, graph theory has provided a powerful and suggestive procedure, to which we turn briefly.

3.2 Graph Theory

If the bonds J_{ij} connecting the N spins S_i are reasonably long-ranged, the nature of the space lattice on which the spins are disposed is of secondary importance, although the dimensionality will, in general, matter. For short-ranged forces the topology of the lattice can play a role, especially for antiferromagnetic couplings. For this reason, we distinguish between the two-dimensional structures: sq (simple quadratic or square), triangular and honeycomb (denoted T and H), and the three-dimensional structures: sc (simple cubic), bcc (body-centered), fcc (face-centered), hexagonal, and other frequently encountered lattice types.

For present purposes, let us extinguish the external field ($B_i = 0$) and allow *only nearest-neighbor* bonds to differ from zero. Assigning to them a strength J, which can be positive (i.e., ferromagnetic) or negative (AF), $w = \tanh\beta J$ becomes the sole parameter in the series expansion (3.1.9). The trace eliminates any single spins. To appear twice, each spin must belong to two distinct, neighboring, bonds. Thus, among the objects contributing to Z_N, there can appear only closed figures —polygons of various sorts, or combinations thereof.

Z_N takes the form

$$Z_N = 2^N (\cosh\beta J)^{\frac{1}{2}Nz} \left[1 + \sum_{r=1}^{\frac{1}{2}Nz} p(r) w^r \right] \quad . \tag{3.2.1}$$

In this expression, z denotes the coordination number of the lattice ($z = 2d$ for the simple lattices in d dimensions such as linear chain, sq, sc, etc.). The total number of links is $\frac{1}{2}Nz$. Also, $p(r)$ is the number of independent closed figures which can be drawn on the given lattice using r links.

In 1D, there are no closed circuits except if we adopt periodic boundary conditions (PBC). For free ends, all $p(r) = 0$. With PBC, all $p(r)$ vanish except for $r = N$, where $p(N) = 1$. Thus the partition function depends on boundary conditions as follows:

1D: $\quad Z_N = 2(2 \cosh\beta J)^{N-1} \quad$ (free ends)

$$= (2 \cosh\beta J)^N + (2 \sinh\beta J)^N \quad \text{(PBC)} \tag{3.2.2}$$

On the two-dimensional sq lattice, $p(1) = p(2) = p(3) = 0$. The lowest order nonvanishing polygon is the square with 4 sides of minimum length. As there are N such polygons, $p(4) = N$. The next nonvanishing order is $r = 6$, of which the 2N contributions are illustrated in Fig.3.1. The vanishing contribution

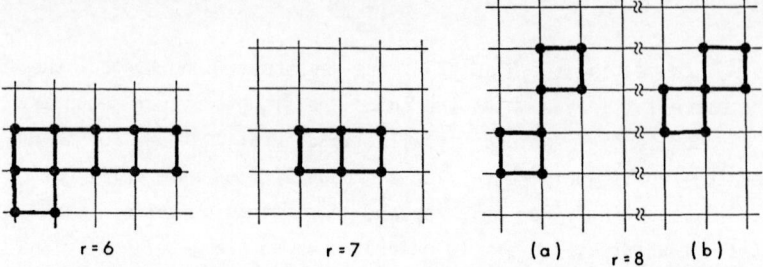

Fig.3.1. Typical diagrams with r = 6 and 8 links which contribute to high-
temperature expansion (3.2.1). The contributions of disconnected loops
(a) differs from that of connected loops (b). A "connection" such as illustrated for r = 7 does not contribute, as shown in Problem 3.4

of r = 7 polygons is the subject of the following problem, thus the next nontrivial contribution comes at r = 8 for which two distinct diagrams must be
considered: disconnected squares which can be placed at $\frac{1}{2}N(N-9)$ locations,
and connected quares, of which there are 2N distinct contributions.

....

Problem 3.4. A 7-sided polygon is illustrated in Fig.3.1. Prove that
p(7) = 0, using the observation that each bond occurs only once, each spin
must appear twice.

....

Thus, the expansion for the sq lattice is:

2D, sq: $Z_N = (2 \cosh\beta J)^{2N}[1 + Nw^4 + 2Nw^6 + 2Nw^8 + \frac{1}{2} N(N - 9)w^8 + \ldots]$

(3.2.3)

a mixed series in powers of w and N. As Z_N must be of the form $(Z_1)^N$, we obtain the w-dependence of $Z_1 = \exp(-\beta f)$, by extracting the Nth root of the
above. In 1D,

1D: $Z_1 = 2 \cosh\beta J$,   (3.2.4)

with both boundary conditions yielding precisely the same result in the thermodynamic limit. In 2D,

2D, sq: $Z_1 = 2 \cosh^2\beta J(1 + w^4 + 2w^6 - 2w^8 + \ldots)$ (3.2.5)

where the coefficients (1,1,2,-2...) are picked so as to fit Z_1^N to Z_N
(3.2.3), to leading orders in N. The same exercise in 3D yields [3.12]:

sc: $Z_1 = 2 \cosh^3\beta J(1 + 3w^4 + 22w^6 + 192w^8 + 2046w^{10} + 24853w^{12} + \ldots)$

bcc: $Z_1 = 2 \cosh^4\beta J(1 + 12w^4 + 148w^6 + 2568w^8 + 53944w^{10} + \ldots)$

$$fcc: \quad Z_1 = 2 \cosh^6 \beta J (1 + 8w^3 + 33w^4 + 168w^5 + 962w^6 + 5928w^7 + 38907w^8$$
$$+ 268056w^9 + \ldots) \quad . \tag{3.2.6}$$

The series are derived for ferromagnets, J (hence w) > 0. For antiferromagnets, we reverse the sign of J, hence of w, finding that only the fcc in (3.2.6) is affected. (The 2D triangular lattice similarly contains odd powers of w, and is affected by a change in sign.) Quite generally, it is found that on *bipartite* lattices (and only on bipartite lattices) the partition function of the ferromagnet (F) and antiferromagnet (AF) are identical in zero external field. We have already encountered this property in connection with the spherical model AF (Sect.2.11). If the lattice can be decomposed into two interpenetrating nets, such that the spins on one sublattice interact only with those on the other, a change in the definition of up and down on one sublattice (but not on the other) effectively changes the sign of J. While the Ising model on a bipartite lattice has a free energy which is an even function of J, the odd powers of J enter in more general cases, including the fcc example above and the triangular lattice illustrated in Fig.3.3. In the figure, it is clear that some of the neighbors of any given spin are themselves neighbors. The consequences for the AF on the triangular and fcc lattices are rather drastic, as we shall ultimately determine. As a preparation, the reader will want to examine the triangular lattice in Problem 3.5.

. . . .

Problem 3.5. Obtain Z_1 to sixth order in w for the triangular lattice, illustrated in Fig.3.3. To this order, plot $\partial^2 Z_1 / \partial w^2$ for $0 < w < 1$ and $0 > w > -1$.

. . . .

We have encountered the lack of symmetry in the sign of the bonds in Sect. 2.7 on long-ranged interactions, where the phenomenon of "frustration" was first examined. As in the long-ranged model, frustration in the triangular AF Ising model destroys the phase transition which exists in the corresponding ferromagnet. The behavior of a quantity analogous to the specific heat is examined in Problem 3.5, serving to illustrate the asymmetrical behavior.

3.3 Low Temperature Expansions and the Duality Relations

Consider the ferromagnetic state, all spins up, as contributing the leading term in an expansion of Z in the number of overturned spins. Overturning a single spin changes the interaction energy with z neighbors from -zJ to

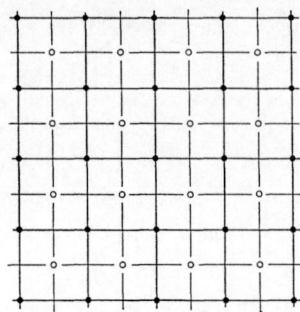

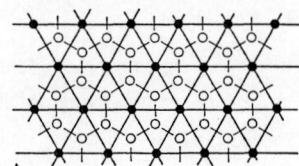

◄ Fig.3.2. Two-dimensional sq net is self-dual

Fig.3.3. Triangular (T) lattice (●●●,━━) is dual of hexagonal (H) lattice (○○○, ──) and vice versa

+zJ — i.e., each bond contributes +2J. The partition function is a power series in $\exp(-\beta 2J)$ but is not simple, especially if overturned spins are neighbors.

In 2D, there exists a geometric relation between any given lattice and a second lattice, denoted the dual lattice. Given the first, the second is obtained by placing points in each elementary cell, connecting them by lines such that each line in the old grid is crossed by a new one. Figure 3.2 shows the sq lattice, which is *self-dual*, and Fig.3.3 the triangular and hexagonal nets, which are duals to one another. We shall see that the low temperature expansion on a given lattice is precisely related to the w expansion on its dual.

Including a factor 2 for the degeneracy (the spins in the ferromagnetic state can equally be all down as up), the low-temperature partition function takes the form

$$Z_N = 2 \exp(\tfrac{1}{2} \beta JzN) \left[1 + \sum_{r \geq 1} \nu(r) \exp(-2r\beta J) \right] \tag{3.3.1}$$

where $\nu(r)$ is the number of configurations having r broken bonds. (A broken bond has its 2 spins antiparallel.)

There is a geometric relation, illustrated in Fig.3.4 for 1,2, and 3 over-turned spins, i.e.: *for any r broken bonds one can draw a closed path of length r on the dual lattice, which crosses every broken bond.* If we denote parameters of the dual lattice by *, we have just proved: $\nu(r) = p^*(r)$ and vice versa, $p(r) = \nu^*(r)$. Thus,

$$Z_N^* = 2^{N^*} (\cosh\beta^* J)^{\frac{1}{2} N^* z^*} \left[1 + \sum_{r \geq 1} p^*(r) w^{*r} \right] \tag{3.3.2}$$

with $w^* = \tanh\beta^* J$. If we pick β and β^* to satisfy the relation,

$$\tanh\beta^* J = \exp(-2\beta J) \tag{3.3.3}$$

we find this relation to be reciprocal, i.e.,

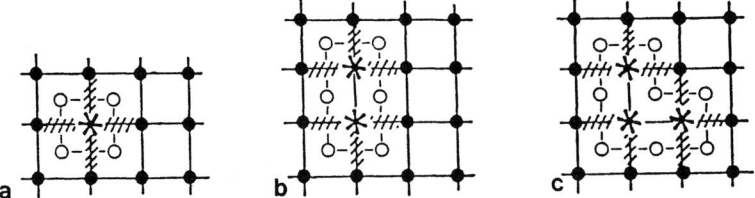

Fig.3.4a-c. Effects of overturning spins. In (a) one overturned spin (*) "breaks" 4 bonds (cross-hatching). A closed path of length r = 4 encloses this spin on the dual lattice. (b) 2 overturned spins break 6 bonds, the shortest path on the dual lattice which crosses these bonds and encloses the 2 spins has length r = 6. (c) Similar for 3 spins, r = 8

$$\tanh\beta J = \exp(-2\beta^* J) \qquad\qquad (3.3.4)$$

and the square brackets in (3.3.1 and 2) to be equal. Therefore,

$$\frac{Z_N^*(\beta^*)}{2^{N^*}(\cosh\beta^* J)^{\frac{1}{2}N^* z^*}} = \frac{Z_N(\beta)}{2\exp(\frac{1}{2}\beta JzN)} . \qquad\qquad (3.3.5)$$

To relate the coordination numbers z and z^* and the lattice numbers N and N^* we need look again at the construction of the dual nets, as in Figs.3.2,3, from which one deduces

$$Nz = N^* z^* \quad\text{and}\quad N + N^* = \frac{1}{2} Nz + 2 . \qquad\qquad (3.3.6)$$

Using these relations together with (3.3.3-5), one proves another useful correspondence:

$$Z_N(\beta)2^{-\frac{1}{2}N}(\cosh 2\beta J)^{-\frac{1}{4}Nz} = Z_N^*(\beta^*)2^{-\frac{1}{2}N^*}(\cosh 2\beta^* J)^{-\frac{1}{4}N^* z^*} \qquad (3.3.7)$$

which is more symmetric than (3.3.5).

Equation (3.3.3 and 4) have the common solution:

$$\sinh 2\beta^* J \sinh 2\beta J = 1 . \qquad\qquad (3.3.8)$$

In the case of the sq lattice which is self-dual, the critical point —if such exists, and if it is unique —must occur at $\beta^* = \beta$. The corresponding T_c ($\sinh 2J/kT_c = 1$) is one of the values given in Table 3.1. In this table, we present T_c on other lattices, by way of comparison. These are obtained by the exact "duality" and "star-triangle" transformations discussed in this chapter for 2D, and by approximate series-expansion methods for 3D and higher.

This table for the Ising ferromagnet may be compared with Table 2.1 for T_c in the spherical model on the hypercubic lattices. Despite a lack of agreement in two dimensions (d = 2) and a discrepancy of some 15% for d = 3,

Table 3.1. T_c for Ising model on various lattices [3.12]

Lattice type	Coordination number (z)	kT_c/zJ
Linear chain	2	0
Simple quadratic (sq)	4	0.567
Simple cubic (sc)	6	0.752
Hypercubic (generalization of the above to d dimensions)	2d $1 \leqslant d \leqslant 4$	$\left(1 - \frac{1}{d}\right)/\ln(1 + \sqrt{2})$
	2d $d \geqslant 4$	$(1 - 0.5962/d)$
Honeycomb	3	0.506
Triangular	6	0.607
Body-centered cubic (bcc)	8	0.794
Face-centered cubic (fcc)	12	0.816

both models ultimately agree with each other and with MFT as $d \to \infty$. The disagreement at $d = 2$ (the spherical model predicts $T_c = 0$ while the duality relation yields a finite value for the Ising model) might occasion some worry, were it not for an independent proof of the existence of a phase transition which we present next.

3.4 Peierls' Proof of Long Range Order

Here, we summarize the arguments by *Peierls* [3.7] subsequently perfected by *Griffiths* [3.8], arguments which can be adapted to all dimensionalities $d \geqslant 2$. They concern the persistence of long-ranged order (LRO) at finite temperature. Calculations based on this procedure yield a lower bound to the critical temperature T_c, just as MFT always provides an upper bound [3.13].

As a preliminary illustration of the line of reasoning, consider first a 1D chain of N spins initially all "up", at $T = 0$. The energetic cost to create a single domain of N_- spins "down" is 2J at each boundary, for a total of 4J. The size of the domain can extend from 1 to N-2 without any change in energy, thus the length of the domain averages $\frac{1}{2} N$ and the average magnetization per spin, the measure of LRO, is zero. Now, if a single domain destroys LRO, the formation of smaller domains (e.g., regions of spins up within the initial domain, regions of spins down within *these*, etc.) will further compound the disorder. At any finite temperature the Boltzmann factor $\exp(-2\beta J)$ for the formation of an individual domain wall is finite, thus the number of such domain walls must be proportional to this Boltzmann factor and to N, the length of the chain. We conclude LRO is surely destroyed at any finite T.

In two dimensions or higher, starting from a situation of all spins up at T = 0, we wish to prove that N_- is a small fraction of N at low temperature, and that the average magnetization per spin $m = 1 - 2N_-/N$ is nonzero. The merging of overturned spins into small *domains*, and the length of domain walls, are essential considerations. For if the overturned spins were isolated, their free energy would be the minimum of the function:

$$F = 8JN_- + kT[N_- \ln N_- + (N - N_-)\ln(N - N_-)] \qquad (3.4.1)$$

on a sq lattice, i.e.,

$$N_- = N(1 + e^{8J/kT})^{-1} , \quad \text{or} \quad m = \tanh 4J/kT \qquad (3.4.2)$$

indicating that the magnetization persists to infinite T. This result is seriously in error, for once the number of overturned spins N_- becomes appreciable, they can no longer be isolated. The energetic cost of adding a spin down to a cluster of overturned spins is significantly less than the energy 8J to overturn a single spin, and must in fact become precisely zero at T_c. [If two neighbors of a given spin are up, and two are down, it no longer has any tendency to be up (or down).]

Peierl's procedure takes into account the variety of shapes of clusters of everturned spins, their energy, and their contribution to the partition function. Because a border of length b encloses at most $b^2/16$ overturned spins, and because we can find an upper bound to the number of such borders n(b) and a lower bound to the energy of each, we can establish a bound on the total number N_- of spins reversed.

First, the total number of possible borders of length b, n(b), satisfies the inequality

$$n(b) \leqslant 4N3^b/3b . \qquad (3.4.3)$$

We *over*estimate the corresponding Boltzmann factor by *under*estimating the cost in energy as +2Jb. Hence, the total number of spins reversed is bounded:

$$N_- \leqslant \sum_{b=4}^{\infty} (b^2/16)(4N3^b/3b)e^{-2Jb/kT}$$

$$\leqslant \frac{Nq^4}{6} \frac{2 - q^2}{(1 - q^2)^2} \qquad (3.4.4)$$

where $q = 3 \exp(-2J/kT)$ must be < 1. Solving this inequality for the value of q which makes $N_- = \frac{1}{2} N$, we can obtain a lower bound on the true T_c at which the spontaneous magnetization disappears. In this way, we find

$$kT_c/4J \geqslant 0.37 \qquad\qquad (3.4.5)$$

or about two-thirds the exact value (Table 3.1, sq. lattice).

When applied to other lattices in 2D and 3D, this procedure again yields reasonable lower bounds on the Curie temperature. *Fisher* has shown that MFT provides upper bounds [3.13], thus $kT_c \leqslant zJ$ with $z = 4$ here.

To complete the proof of the existence of a phase transition, one must show that the spins are *uncorrelated* at sufficiently high temperature. One simple argument states that as all thermodynamic functions depend on the ratio J/kT only, the limit $T = \infty$ is the same as $J = 0$; there we know that all spins are uncorrelated. Quantitatively, to evaluate $\langle S_n S_m \rangle_{TA}$ we can use (3.1.8):

$$\langle S_n S_m \rangle_{TA} = \frac{1}{Z_N}\, \partial^2 Z_N/\partial w_n \partial w_m \bigg|_{\substack{all \\ w_j=0}} . \qquad\qquad (3.4.6)$$

The so-called magnetic graphs which contribute in the above start at n and terminate at m, thus involve L_{nm} products of the nearest-neighbor functions w_{ij}, where L_{nm} is the number of links separating n from m. At high temperatures, $w_{ij} \simeq J/kT \to 0$, so we expect *an exponential decrease in the correlation function* such as

$$\langle S_n S_m \rangle_{TA} \sim \exp(-L_{nm}/\xi) \qquad\qquad (3.4.7)$$

with the correlation length ξ a function of T. Evidently, $\xi \to \infty$ at T_c with the onset of long range order (LRO); assuming a power-law behavior, one defines a new critical index ν:

$$\xi = \left(\frac{T - T_c}{T_c}\right)^{-\nu} . \qquad\qquad (3.4.8)$$

The divergence of ξ is closely related to the divergence in the susceptibility. (The left-hand side of (3.4.7) is just the summand in the high-temperature susceptibility, (2.2.15).)

3.5 1D Ising Model in Longitudinal Fields

In this section we derive the exact solution of the 1D Ising model with nearest-neighbor interactions, including such parallel "longitudinal" interactions as an external magnetic field parallel to the axis of quantization, and interactions with the underlying normal modes (phonons) of the molecule or solid to which the spins are pinned. We leave to the following section the study of transverse fields —those that involve operators such as S_i^x or S_i^y which do not commute with the S_i^z of the Ising model. (The *combination* of transverse *and* longitudinal external fields has not been solved in closed form.) Although the model is classical —all the operators commute, all spatial configurations can be specified without uncertainty —the study of the associated statistical mechanics quickly involves us in algebra of operators which, in general, do not commute with one another.

A reasonably general starting point is provided by the Hamiltonian,

$$H = - \sum_{n=1}^{N} J_n S_n S_{n+1} - \sum_{n=1}^{N} B_n S_n \quad . \tag{3.5.1}$$

We may write this as $\sum H_n$, and associate with each H_n a Boltzmann factor V_n:

$$V_n = \exp\left[\beta J_n S_n S_{n+1} + \frac{1}{2} \beta (B_n S_n + B_{n+1} S_{n+1})\right] \tag{3.5.2}$$

with some allowance at $n=1$ and N for desired boundary conditions.

The partition function is then

$$Z_N = \text{Tr}\{\ldots V_n V_{n+1} \ldots\} \quad . \tag{3.5.3}$$

There is some advantage in considering the V_n's to be matrices of the following form:

$$V_n \rightarrow V(S_n; S_{n+1}) \tag{3.5.4}$$

i.e., 2×2 matrices in the indices $S_n = \pm 1$, and $S_{n+1} = \pm 1$. Matrix multiplication takes the form

$$V(S_{n-1}; S_n) \cdot V(S_n; S_{n+1}) \equiv \sum_{S_n=-1}^{+1} V(S_{n-1}; S_n) V(S_n; S_{n+1})$$

$$\equiv W(S_{n-1}; S_{n+1}) \tag{3.5.5}$$

where W is the resultant product matrix, also 2×2.

The advantage lies in the fact that *the operation "Tr" is equivalent to* $\prod_n (\sum_{S_n})$, *i.e. to repeated matrix multiplication.*

So, according to (3.5.5),

$$Z_N = \text{Tr}\{V_1 \cdot V_2 \ldots V_n \cdot V_{n+1} \ldots V_N\}$$

$$= \text{tr}\{U(S_1; S_N)\} \tag{3.5.6}$$

where the last operation, tr, is the sum of the diagonal elements of the 2×2 matrix

$$U = V_1 \cdots V_n \cdot V_{n+1} \cdots V_N \quad . \tag{3.5.7}$$

Explicitly, each V_n matrix takes the form:

$\searrow$ $\overset{\rightarrow}{S_{n+1}}$ $\downarrow$ S_n	+1	-1
+1	$\exp\left[\beta J_n + \frac{1}{2}\beta(B_n + B_{n+1})\right]$	$\exp\left[-\beta J_n + \frac{1}{2}\beta(B_n - B_{n+1})\right]$
-1	$\exp\left[-\beta J_n - \frac{1}{2}\beta(B_n - B_{n+1})\right]$	$\exp\left[\beta J_n - \frac{1}{2}\beta(B_n + B_{n+1})\right]$

$V_n =$

$$\tag{3.5.8}$$

In terms of the Pauli spin matrices

$$1 = \begin{bmatrix} 1 & 0 \\ 0 & 1 \end{bmatrix}, \ \sigma_x = \begin{bmatrix} 0 & 1 \\ 1 & 0 \end{bmatrix}, \ \sigma_y = i\begin{bmatrix} 0 & -1 \\ 1 & 0 \end{bmatrix}, \ \text{and} \ \sigma_z = \begin{bmatrix} 1 & 0 \\ 0 & -1 \end{bmatrix}$$

(3.5.8) can be written compactly

$$V_n = \exp(\beta J_n) \cosh \frac{1}{2}\beta(B_n + B_{n+1})1 + \exp(\beta J_n) \sinh \frac{1}{2}\beta(B_n + B_{n+1})\sigma_z$$

$$+ \exp(-\beta J_n) \cosh \frac{1}{2}\beta(B_n - B_{n+1})\sigma_x + i \exp(-\beta J_n) \sinh \frac{1}{2}\beta(B_n - B_{n+1})\sigma_y \quad .$$

$$\tag{3.5.9}$$

As we have stated before, V_1 and V_N are special cases. The manner in which we treat them does not influence extensive quantities, but is important in boundary phenomena such as surface tension.

In the case of constant nearest-neighbor interactions $J_n = 1$ and homogeneous field $B_n = B$, the V matrices are all identical. With periodic boundary conditions, the U in (3.5.6) becomes

$$U = V^N \quad , \tag{3.5.10}$$

and therefore

$$Z_N = \lambda_+^N + \lambda_-^N \tag{3.5.11}$$

where $\lambda_\pm$ are the two eigenvalues of V. We solve for them in the usual way:

$$\det \left\| \begin{array}{cc} K_1 K_2 - \lambda & K_1^{-1} \\ K_1^{-1} & K_1 K_2^{-1} - \lambda \end{array} \right\| = 0 \tag{3.5.12}$$

104

with $K_1 \equiv \exp(\beta J)$ and $K_2 \equiv \exp(\beta B)$. The solution is

$$\lambda_\pm = \frac{1}{2} K_1 (K_2 + K_2^{-1}) \pm \left[\frac{1}{4} K_1^2 (K_2 + K_2^{-1})^2 - (K_1^2 - K_1^{-2}) \right]^{\frac{1}{2}}$$

$$= e^{\beta J} \cosh\beta B \pm (e^{2\beta J} \sinh^2\beta B + e^{-2\beta J})^{\frac{1}{2}} . \qquad (3.5.13)$$

Evidently, λ_+ exceeds λ_-. When each is raised to the Nth power and N becomes large, we find

$$Z_N = \lambda_+^N [1 + (\lambda_-/\lambda_+)^N] \rightarrow \lambda_+^N \qquad (3.5.14)$$

as the ratio $(\lambda_-/\lambda_+)^N$ becomes exponentially small in the limit of large N.

The above illustrates a general principle in calculating Z_N. $\mathbf{V}$ is denoted the *transfer matrix* and λ_+ is its largest or *optimal* eigenvalue. *We need obtain only the largest eigenvalue of the transfer matrix*, the contribution of the other(s) vanishing at large N. This is a consequence of the criterion developed in Chap.2, the minimization of the free energy.

The Nth root of (3.5.14) is trivially

$$Z_1 = \lambda_+ \qquad (3.5.15)$$

with λ_+ given in (3.5.13). For zero external field, this yields precise agreement with (3.2.2,4) found by series expansion in zero field. In vanishing field, we now obtain the specific heat and paramagnetic susceptibility:

$$c = k(\beta J)^2 / \cosh^2 \beta J \qquad (3.5.16)$$

and

$$\chi_0 = \frac{\mathbb{C}_{\frac{1}{2}}}{T} e^{2\beta J} . \qquad (3.5.17)$$

The latter *diverges* at T = 0 for ferromagnetic coupling (J > 0), and *vanishes* exponentially for antiferromagnetic coupling (J < 0) at low temperature, a common occurrence in 1D, as previously discussed [see (2.16.24)ff., and (2.16.25) relating ferromagnetic to antiferromagnetic susceptibility]. At high T,

$$\chi_0^{-1} \propto T[1 - 2\beta J + O(\beta J)^2] = T - 2J/k + O(1/T) . \qquad (3.5.18)$$

While this is precisely the Curie-Weiss law, as usual this expansion is misleading near the critical point, T = 0 in this instance. Equations (3.5.16-18) are illustrated in Fig.3.5, which indicates the high-temperature extrapolation.

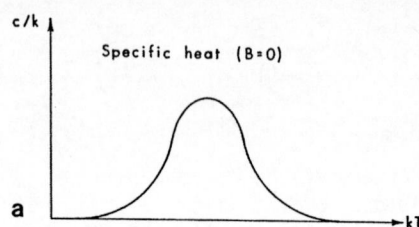
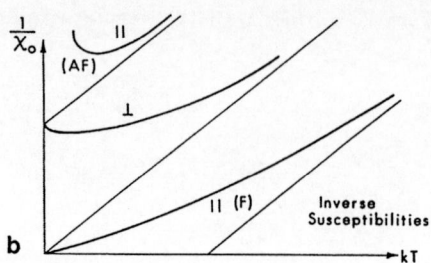

Fig.3.5a,b. Properties of 1D Ising model: Specific heat in zero field, and various zero-field susceptibilities. $1/\chi_0$ is plotted to show asymptotic approach to Curie-Weiss laws for antiferromagnet (AF) and ferromagnet (F), and are labelled ∥ to indicate magnetic field is applied parallel to spins. The perpendicular susceptibility (⊥) is the same for F and AF, see (3.6.27)

In zero external field the transfer matrices all commute and can be diagonalized simultaneously. Consequently, we obtain Z for an *arbitrary* distribution of J_n

$$Z_N = \prod_n 2 \cosh\beta J_n \quad ,$$

or better

$$f = - \frac{kT}{N} \sum_n \ln(2 \cosh\beta J_n) = -kT\langle\ln(2 \cosh\beta J)\rangle \tag{3.5.19}$$

where the average is over the given distribution of J_n's. We can treat the case of "random bonds" similarly, that is, the case when the bond strengths are given only by a statistical distribution.

Given arbitrary bonds *and* and external field, the evaluation of the partition function becomes more difficult as, in general, the matrices (3.5.9) no longer commute with one another. Let us demonstrate the procedure when first one transfer matrix differs from the rest and later, when two such matrices are distinguished. Let the pth matrix in (3.5.6) be $V' \neq V$. Only the eigenvector u_+ associated with λ_+ contributes, therefore we only need calculate:

$$Z_N = Z_N^0 \frac{u_+^T \cdot V^{p-1} \cdot V' \cdot V^{N-p} \cdot u_+}{u_+^T \cdot V^{p-1} \cdot V \cdot V^{N-p} \cdot u_+} \tag{3.5.20}$$

Here, Z_N^0 is the partition function when all the matrices are equal, and the eigenvectors (and transposes) are

$$V \cdot u_+ = \lambda_+ u_+ \quad , \quad u_+^T \cdot V = \lambda_+ u_+^T \tag{3.5.21}$$

with similar expressions for u_-, the second eigenvector (associated with λ_-). Using the eigenvalue relation above, we reduce (3.5.20) to a ground state expectation value

$$Z_N = Z_N^0 \frac{1}{\lambda_+} u_+^T \cdot V' \cdot u_+ \quad . \tag{3.5.22}$$

In terms of free energy

$$F = F^0 + kT[\ln\lambda_+ - \ln(u_+^T \cdot V' \cdot u_+)] \quad . \tag{3.5.23}$$

The case of *two* different V's is more illuminating. Assume they are separated by n host or unperturbed transfer matrices. We then find

$$Z_N = Z_N^0 \frac{u_+^T \cdot V' \cdot V^n \cdot V'' \cdot u_+}{u_+^T \cdot V^{n+2} \cdot u_+}$$

$$= Z_N^0 \lambda_+^{-2} \Big[(u_+^T \cdot V' \cdot u_+)(u_+^T \cdot V'' \cdot u_+) + (\lambda_-/\lambda_+)^n (u_+^T \cdot V' \cdot u_-)(u_-^T \cdot V'' \cdot u_+) \Big] \quad . \tag{3.5.24}$$

This is obtained by using *completeness* $u_+u_+^T + u_-u_-^T = 1$ twice: once following V', and once preceding V''. Taking logarithms, we find that this expression contains the free energy of the individual defects as in (3.5.23) plus a two-defect interaction term which involves u_- and λ_-.

This expression also suggests how one is to obtain *correlation functions* in the homogeneous chain (all $J_n = J$, all $B_n = B$). Suppose we wished to obtain the thermal averaged correlation function of S_p and S_{p+n+1}, i.e., the two spins separated by n intervening sites. We define two matrices V' and V'' as follows:

$$V' = V(S_{p-1};S_p)S_p \quad \text{and} \quad V'' = V(S_{p+n};S_{p+n+1})S_{p+n+1} \quad . \tag{3.5.25}$$

. . . .

Problem 3.6. Using the above, derive the 2×2 matrix σ in the following relations:

$$m \equiv <S_p>_{TA} = u_+^T \cdot \sigma \cdot u_+ \quad \text{and} \tag{3.5.26}$$

$$<(S_p - m)(S_{p+n+1} - m)>_{TA} = \left(\frac{\lambda_-}{\lambda_+}\right)^{n+1} (u_+^T \cdot \sigma \cdot u_-)(u_-^T \cdot \sigma' \cdot u_+) \quad , \tag{3.5.27}$$

and show that the first agrees with $-\partial F/\partial B$ calculated from (3.5.14). Calculate (3.5.27) in zero field ($B = 0$) and verify that it yields the bond energy at $n = 0$, as well as χ_0, (3.5.17), summing all n.

. . . .

The result (3.5.27) leads to two significant observations. First, that the excited spectrum of the transfer matrix (λ_- in this instance) is needed to calculate the correlation functions. Second, the correlations decay exponentially, with a "correlation length" $\xi \sim 1/\ln(\lambda_+/\lambda_-)$. For thermodynamic functions (Z,m,etc.) only λ_+ is needed. But in observing the properties of a system, e.g. by neutron scattering where the correlation functions are invaluable, we should have the complete spectrum λ's and u's of V.

In practice, J_n could exhibit some variability due to *magnetostriction*. This rather general phenomenon comes about as the result of coupling between the magnetic and mechanical degrees of freedom of the lattice, and is particularly easy to analyze in one dimension. Suppose the ideal interatomic distances are a, and that the exchange bonds J_n vary with *actual* interatomic distance. Expanding about equilibrium value a:

$$J(x_{n+1} - x_n) = J(0)\left[1 + n_1 \frac{(x_{n+1} - x_n)}{a} + n_2 \frac{(x_{n+1} - x_n)^2}{a^2} + \dots\right] .$$

$$(3.5.28)$$

The total Hamiltonian includes

$$H_{phonon} = \sum_n \left[p_n^2/2M + \frac{1}{2} M\omega_0^2(x_{n+1} - x_n)^2\right]$$

$$(3.5.29)$$

$$H_{Ising} = -J(0) \sum_n S_n S_{n+1} - \sum_n B_n S_n$$

comprising the unperturbed Hamiltonian. The first-order coupling is

$$H_1 = -\frac{J(0)}{a} n_1 \sum_n (x_{n+1} - x_n)S_n S_{n+1} ,$$

$$(3.5.30)$$

and the second-order coupling is

$$H_2 = -\frac{J(0)}{a^2} n_2 \sum_n (x_{n+1} - x_n)^2 S_n S_{n+1} .$$

$$(3.5.31)$$

The effects of the first-order coupling can be derived without further approximation.

The method is straightforward. We merely shift the origin of each atomic coordinate by an amount which depends on the magnetic configuration. Thus, with

$$x_n \rightarrow x_n + \frac{J(0) n_1}{M\omega_0^2 a} \sum_{j<n} S_j S_{j+1}$$

$$(3.5.32)$$

we eliminate H_1 and obtain

$$H = H_{ph} + H_I - \frac{1}{2} N \frac{J^2(0)\eta_1^2}{M\omega_0^2 a^2} + H_2 \quad . \tag{3.5.33}$$

The effect of H_2 can be estimated (Problem 3.7). According to (3.5.32), the main contribution of η_1 is to the length of the chain — hence the term magnetostriction. If $<x_N - x_0> + Na$ was initially L, we have now (for $B = 0$):

$$L \rightarrow L - (\eta_1/M\omega_0^2 a)U(T) \tag{3.5.34}$$

where $U(T)$ is the internal energy of the Ising Hamiltonian with $B = 0$. (In finite field, the expression is somewhat more cumbersome.) The sign of the magnetostrictive effect depends on the sign of η_1. The result, $\partial L/\partial T \propto -\eta_1 c(T)$ agrees with experiment even in three dimensions — even in MFT [3.14].

....

Problem 3.7. Suppose $\eta_2 = 0(\eta_1^2)$, and treat H_2 in leading order. Let

$$H_2 \approx - \frac{J(0)}{a^2} \eta_2 \sum_n [<(x_{n+1} - x_n)^2>_{TA} S_n S_{n+1} + (x_{n+1} - x_n)^2 <S_n S_{n+1}>_{TA}$$

$$- <(x_{n+1} - x_n)^2>_{TA} <S_n S_{n+1}>_{TA}]$$

and calculate the effects of each term to leading order, obtaining quantities such as new speed of sound, new J, etc. [3.15].

....

3.6   1D Ising Model in Transverse Fields

Until now, the spins and the external and internal forces acting on them have all lain along a preferred axis. We here consider situations in which the two-body forces involve components of the spins along one axis, while the external field interacts with spin components along a perpendicular axis. We observed that in MFT, the susceptibility of the antiferromagnet is quite different when the field is transverse rather than longitudinal, the discrepancies being highlighted in Fig.2.8. In this section, we shall corroborate this in the context of a microscopic theory. Our calculations will also serve for the statistical mechanics of the *two*-dimensional Ising model considered later on this chapter.

The Hamiltonian now contains non-commuting operators S_i^z and S_i^x, for example:

$$H = -J \sum_n S_n^z S_{n+1}^z - B \sum_n S_n^x \quad . \tag{3.6.1}$$

This describes a simple, homogeneous case: constant nearest-neighbor bonds and a constant external field. Nevertheless, we can no longer write

$$\exp(-\beta H) = \exp(-\beta H_1) \exp(-\beta H_2) \ldots$$

as in (3.5.2-6) because the H_n's do not commute with their neighbors $H_{n\pm1}$. We are thus forced to seek an alternative to the transfer matrix approach, a complication typical of *quantum* statistical mechanics, generally requiring one to diagonalize the complete Hamiltonian before attempting to evaluate Z_N. In this instance we process somewhat more information prior to such a calculation, thanks to a "*duality*" transformation which maps J and B into each other. The details follow.

Let us write S_n^z and S_n^x in terms of products of other Pauli operators T_m^z and T_m^x as follows:

$$S_n^z = \prod_{m=1}^{n} T_m^x \quad \text{and} \quad S_n^x = T_n^z T_{n+1}^z \tag{3.6.2}$$

....

Problem 3.8. Show that

$$S_n^y = -T_n^y T_{n+1}^z \prod_{m=1}^{n-1} T_m^x . \tag{3.6.3}$$

Prove that the $\{S_n\}$ satisfy the correct Pauli matrix algebra, including the fact that commutators of spins on different sites vanish.

....

Forthwith, (3.6.1) takes the form

$$H = -B \sum_n T_n^z T_{n+1}^z - J \sum_n T_n^x \tag{3.6.4}$$

which is precisely the same, with J,B interchanged. We can infer that if there are only two phases- one corresponding to large $|J/B|$ and the other to small $|J/B|$, the critical point must lie at $|J/B| = 1$. Quantities such as the ground state energy will be symmetric about that point.

For definiteness, the following calculations are based on (3.6.1), although they could just as easily apply to (3.6.4).

Equation (3.6.1) under consideration contains both quadratic and linear expressions in the spins. Unlike similar expressions in the Gaussian and spherical models, these cannot be reduced to diagonal form (a quadratic form in the spin operators in which the various normal modes are decoupled). The principal reason lies with the commutation relations. Whereas Pauli matrices on a given site satisfy anticommutation relations, matrices associated with two different spins commute with one another. Linear combinations of spin operators neither commute nor anticommute with one another — their algebraic properties are highly entangled.

As a remedy, we shall seek to construct a set of operators that anticommute everywhere. The transformation from Pauli operators to fermions by means of the Jordan-Wigner transformation was already examined in [Ref.3.11, Chap.3], and is entirely suitable to this purpose.

We introduce a set of anticommuting operators c_j and c_j^*, and a number operator $n_j = c_j^* c_j$ the eigenvalues of which are 0,1. The spins can be expressed entirely with the aid of these operators, in the form

$$S_j^x = S_j^+ + S_j^- \quad , \quad \text{where} \quad S_j^- = c_j \exp\left(i\pi \sum_{r<j} n_r\right) \quad \text{and} \quad S_j^+ = (S_j^-)^+$$

$$\text{and} \quad S_j^z = 2n_j - 1 \quad . \tag{3.6.5}$$

On a given site, the anticommutation relations of c_j and c_j^* are the same as for Pauli matrices; the exponential phase factors cancel out. On two different sites, these phase factors change a commutator into an *anti*commutator. In this sense, this transformation is *not* unitary, it does not preserve commutation relations. Nevertheless, it is well defined and leads to an ordinary eigenvalue problem for spins 1/2 with nearest-neighbor interactions on the linear chain. Consider the quadratic form $S_j^x S_{j+p}^x$:

$$(S_j^+ S_{j+p}^+ + S_j^+ S_{j+p}^-) + \text{H.C.} \tag{3.6.6}$$

where H.C. stands for Hermitean conjugate also indicated by $^+$ as in (3.6.5). In terms of Fermi operators:

$$S_j^+ S_{j+p}^+ = c_j^* c_{j+p}^* \exp\left(i\pi \sum_{j \leqslant r < j+p} n_r\right)$$

$$S_j^+ S_{j+p}^- = c_j^* c_{j+p} \exp\left(i\pi \sum_{j \leqslant r < j+p} n_r\right) \quad . \tag{3.6.7}$$

It is useful to note the identity,

$$\exp(i\pi n_r) \equiv 1 - 2n_r \equiv (c_r^* + c_r)(c_r^* - c_r) \tag{3.6.8}$$

and that the square of each side is 1. Further,

$$c_r^* \exp(i\pi n_r) = c_r^* \quad , \quad c_r \exp(i\pi n_r) = -c_r \quad . \tag{3.6.9}$$

Thus, for $p = 1$ in (3.6.7) we obtain simply

$$S_j^+ S_{j+1}^+ = c_j^* c_{j+1}^* \quad \text{and}$$

$$S_j^+ S_{j+1}^- = c_j^* c_{j+1} \quad . \tag{3.6.10}$$

For $p = 2$, the expressions are

$$S_j^+ S_{j+2}^+ = c_j^* c_{j+2}^* (1 - 2n_{j+1})$$

$$S_j^+ S_{j+2}^- = c_j^* c_{j+2} (1 - 2n_{j+1}) \quad . \tag{3.6.11}$$

As the number operator $n_{j+1} = c_{j+1}^* c_{j+1}$ is itself bilinear in fermion field operators, we see that for $p = 1$ the expressions are quadratic, for $p = 2$ they are quartic, etc.

Fortunately, because the starting Hamiltonian (3.6.1) involves *only* nearest-neighbor bonds, $p = 1$. As it is written, with the bonds in terms of S_j^z operators, the transformation is not of much use. But an initial rotation of $90°$ in spin space: $S_j^x \to -S_j^z$, $S_j^z \to +S_j^x$ for all j, transforms H into the form:

$$H \to -J \sum_j S_j^x S_{j+1}^x + B \sum_j S_j^z \tag{3.6.12}$$

This can be rendered uniformly quadratic in fermions by (3.6.5):

$$H \to -J \sum_j (c_j^* c_{j+1}^* + c_j^* c_{j+1} + c_{j+1}^* c_j + c_{j+1} c_j) + B \sum_j (2c_j^* c_j - 1) \quad , \tag{3.6.13}$$

in which form it can be easily diagonalized after Fourier transformation. In the duality transformation, and in the Jordan-Wigner transformation to fermions we have neglected end effects (at $j = 1$ or N). We continue to do this, for simplicity in obtaining the normal modes of (3.6.13). Assuming translational invariance (PBC in the present form of H) one starts by Fourier transforming, introducing a new complete set of fermion operators a_k and a_k^* [$k = (2\pi/N) \cdot$ integer] by what amounts to a unitary transformation from the set of c's to the set of a's.

$$c_n = N^{-\frac{1}{2}} \sum_k a_k e^{ikn} \quad \text{and}$$

$$c_n^* = N^{-\frac{1}{2}} \sum_k a_k^* e^{-ikn} \tag{3.6.14}$$

where sums over k range over the N equally spaced points in the interval $-\pi < k \leq +\pi$. Like the c_n's, the a_k's also satisfy fermion anticommutation relations:

$$a_k^* a_{k'}^* + a_{k'}^* a_k^* \equiv \{a_k^*, a_{k'}^*\} = 0 = \{a_{k'}, a_k\} \qquad \text{and}$$

$$\{a_k^*, a_{k'}\} = \delta_{k,k'} \quad . \tag{3.6.15}$$

That *such* algebraic bilinear relations are preserved by a unitary transformation (3.6.14), is verified in Problem 3.9.

....

Problem 3.9. Using (3.6.14,15) above to *define* the c_n's and c_n^*'s, verify

$$c_j^2 = c_j^{*2} = 0 \quad , \quad \{c_j^*, c_r\} = \delta_{jr}$$

and finally, prove (3.6.9).

....

We insert (3.6.14) into H, (3.6.13), obtaining:

$$H = -JN^{-1} \sum_j \sum_k \sum_{k'} e^{i(k-k')j} \left[a_k^* a_{-k}^* e^{ik'} + a_k^* a_{k'} (e^{ik'} + e^{-ik}) + a_{-k} a_k e^{-ik} \right]$$

$$+ 2BN^{-1} \sum_j \sum_k \sum_{k'} e^{i(k'-k)j} a_k^* a_{k'} - NB \quad . \tag{3.6.16}$$

The orthonormality relations which appear in this context

$$\frac{1}{N} \sum_j e^{i(k'-k)j} = \delta_{k,k'} \tag{3.6.17}$$

are tantamount to conservation of momentum. With them, we eliminate the sums over j and k' and bring H to the form $\sum H_k$,

$$H = -J \sum_{\pi > k > 0} \left[(a_k^* a_{-k}^* - a_{-k} a_k)(2i \sin k) + (a_k^* a_k + a_{-k}^* a_{-k})(2 \cos k) \right]$$

$$+ 2B \sum_{\pi > k > 0} (a_k^* a_k + a_{-k}^* a_{-k}) + H_0 + H_\pi - NB \quad . \tag{3.6.18}$$

The individual H_k now commute with one another and can be individually diagonalized. At $k = 0$ or π, they are *already* diagonal:

$$H_0 = -2(J - B) a_0^* a_0 \qquad \text{and}$$

$$\tag{3.6.19}$$

$$H_\pi = 2(J + B) a_\pi^* a_\pi \quad .$$

The $k = 0$ mode is significant for ferromagnetism $(J > 0)$ whereas $k = \pi$ is important *for* antiferromagnetism $(J < 0)$. In either case we see confirmation of the critical point at $|J/B| = 1$, with the occupancy of the relevant mode changing from 0 to 1.

But, whatever the behavior of these two particular modes, they have no bearing on the *total* energies which are O(N). We must diagonalize *all* the H_k, then sum over their ground state energies (at $T = 0$) or over their free energies to obtain the system properties. We write H_k in the form

$$H_k = -J(a_k^* a_{-k}^* - \text{H.C.})2i \sin k + 2(B - J \cos k)(a_k^* a_k + a_{-k}^* a_{-k} - 1)$$

$$(3.6.20)$$

after adding and subtracting $-2J \sum_{\pi > k > 0} \cos k = 0$ to H in (3.6.18).

The operators a^* occur only in pairs $a_k^* a_{-k}^*$, so H_k connects only even-occupancy states with each other, or odd-occupancy states with each other. Thus, $|0)$ is connected by H_k to $a_k^* a_{-k}^* |0)$ but not to $a_k^* |0)$ or $a_{-k}^* |0)$. In fact, the last two are —by inspection— individual eigenstates of H_k with eigenvalues zero. Within the even-occupancy subspace, H_k takes the form:

$$H_k = -2J \sin k \ S_k^y + 2(B - J \cos k)S_k^z \qquad (3.6.21)$$

where the S^y and S^z matrices are the usual Pauli matrices in the space of:

$$|0) = \begin{bmatrix} 0 \\ 1 \end{bmatrix} \quad \text{and} \quad a_k^* a_{-k}^* |0) = \begin{bmatrix} 1 \\ 0 \end{bmatrix} . \qquad (3.6.22)$$

This is the Hamiltonian of a single "spin" in an external field tilted in the y,z plane. The eigenvalues are $\pm \Delta_k$, the magnitude of the external field:

$$\Delta_k \equiv 2[(J \sin k)^2 + (B - J \cos k)^2]^{\frac{1}{2}} . \qquad (3.6.23)$$

The ground state is always $-\Delta_k$; the remaining eigenvalues are 0,0 and $+\Delta_k$; however, all four must be taken into account in evaluating the free energy. For F_k we find

$$F_k = -kT \ \ln\{4 \cosh^2 \beta[(J \sin k)^2 + (B - J \cos k)^2]^{\frac{1}{2}}\} \qquad (3.6.24)$$

after use of a half-angle trigonometric identity. The total free energy is thus

$$F = -kT \ N \ \frac{1}{2\pi} \int_{-\pi}^{+\pi} dk \ \ln\{2 \cosh\beta[\ ]^{\frac{1}{2}}\} \qquad (3.6.25)$$

with $[\ ] = 1/2\Delta_k$ as in (3.6.24) above.

114

Similarly, the ground state energy is

$$E_0 = -\frac{N}{2\pi} \int_{-\pi}^{+\pi} dk[\ ]^{\frac{1}{2}} \ . \tag{3.6.26}$$

Without further analysis, we obtain the transverse magnetic susceptibility $\chi_\perp$ in vanishing field. Differentiation of F twice and the setting of B = 0 . results in trivial integrations, with the result:

$$\chi_\perp = -\mathfrak{C}_{\frac{1}{2}} \left.\frac{\partial^2 F}{\partial B^2}\right|_{B=0} = \frac{1}{2} N\mathfrak{C}_{\frac{1}{2}}[J^{-1} \tanh\beta J + \beta\cosh^{-2}\beta J] \ . \tag{3.6.27}$$

This quantity differs from the longitudinal susceptibility, calculated earlier, in being symmetric in J. It thus extrapolates to a pure Curie law at high T, not to a Curie-Weiss law. It also is finite at T = 0, unlike the longitudinal susceptibility which is either 0 (AF) or ∞ (ferromagnetic couplings). These various results are compared graphically in Fig.3.5 of the preceding section.

At finite B the results are considerably more involved. F must be evaluated numerically, while E_0 and its derivatives are expressible only in terms of complete elliptic functions. These are tabulated and their properties near critical points are well known, but we can obtain a qualitative understanding more simply by analyzing the spectrum Δ_k of elementary excitations.

This spectrum is given by Δ_k (the lowest excitation energy is from the ground state of any H_k to either of the odd-occupation states having energy zero). For J > 0 and B = 0 these excitations are easily interpreted as the energy 2J to create one additional *domain wall* in the system (taking any two parallel spins, and overturning the second member of this pair together with all the spins to the right of it). As B is increased, the geometric interpretation becomes more difficult. But we note that with increasing B, the gap against the elementary excitations decreases until, at B = J it vanishes altogether. Indeed, the symmetry between B and J which was revealed through the duality transformation (3.6.2) is also reflected in the spectrum of Δ_k, which we here rewrite in symmetric form:

$$\Delta_k = 2[J^2 + B^2] \left(1 - \frac{2BJ}{B^2 + J^2} \cos k\right)^{\frac{1}{2}}$$

$$= 2(|J| + |B|)\left(1 - \frac{4|BJ|}{(|J| + |B|)^2} g_k\right)^{\frac{1}{2}} \tag{3.6.28}$$

where $g_k = \cos^2 \frac{1}{2} k$ for BJ > 0 and $\sin^2 \frac{1}{2} k$ for BJ < 0.

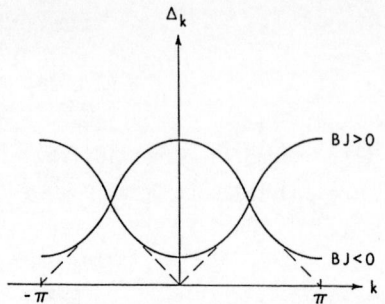

Fig.3.6. Excitation spectrum of 1D Ising model (coupling constant J) in transverse magnetic field B. Spectra for BJ > 0 and < 0 are shifted by π with respect to each other. The spectra can go to zero for the critical values $|B/J| = 1$ (----)

This spectrum is plotted in Fig.3.6 for the general cases $B \neq J$ as well as in the singular cases $|B| = |J|$ for which the gap $\|J| - B|$ vanishes.

To evaluate (3.6.26) and its derivatives, we invoke the complete elliptic integrals [3.16]:

$$E(k) = \int_0^{\frac{1}{2}\pi} d\phi (1 - k^2 \sin^2\phi)^{\frac{1}{2}} \quad \text{and}$$

$$K(k) = \int_0^{\frac{1}{2}\pi} d\phi (1 - k^2 \sin^2\phi)^{-\frac{1}{2}} \quad .$$

(3.6.29)

Thus, (3.6.26) is obtained in closed form:

$$E_0 = -N \frac{|J| + |B|}{\pi} E\left(\frac{2|BJ|^{\frac{1}{2}}}{|J| + |B|} \right) \quad .$$

(3.6.30)

The singular behavior near the critical point is best analyzed with the aid of asymptotic expansions developed for this purpose [3.16]. Let $k'^2 = 1 - k^2$, i.e., with

$$k'^2 = \frac{(|J| - |B|)^2}{(|J| + |B|)^2} \quad \text{and} \quad \Lambda \equiv \ln|4/k'| \quad ,$$

one finds

$$E = 1 + \frac{1}{2} (\Lambda - \frac{1}{2})k'^2 + \frac{3}{16} (\Lambda - \frac{13}{12})k'^4 + \dots$$

and

(3.6.31)

$$K = \Lambda + \frac{1}{4} (\Lambda - 1)k'^2 + \dots \quad .$$

Thus, while K diverges logarithmically at $k' = 0$, it is only the second derivative of E, $\partial^2 E/\partial k'^2$ which is logarithmically divergent at that point.

These results will be relevant to the two-dimensional Ising model, which is reducible to the 1D model in a transverse field.

There exists a more direct method for diagonalizing the H_k of (3.6.20). It employs a unitary transformation due to Bogolubov, which found many applications in the Bardeen-Cooper-Schrieffer theory of superconductivity. If we describe the transformation as acting on arbitrary operators in the following manner:

$$\text{op} \rightarrow \exp\left[-\theta_k(a_k^* a_{-k}^* - a_{-k}a_k)\right]\left[\exp \tfrac{1}{4} i\pi(n_k + n_{-k})\right] \text{op} \exp\left[-\tfrac{1}{4} i\pi(n_k + n_{-k})\right]$$

$$\times \exp\left[\theta_k(a_k^* a_{-k}^* - a_{-k}a_k)\right] \qquad (3.6.32a)$$

then it specifically transforms the fermion field operators as follows,

$$a_k \rightarrow e^{-\frac{1}{4}i\pi}(a_k \cos\theta_k + a_{-k}^* \sin\theta_k)$$

$$a_{-k} \rightarrow e^{-\frac{1}{4}i\pi}(a_{-k} \cos\theta_k - a_k^* \sin\theta_k)$$

$$a_k^* \rightarrow e^{\frac{1}{4}i\pi}(a_k^* \cos\theta_k + a_{-k} \sin\theta_k)$$

$$a_{-k}^* \rightarrow e^{\frac{1}{4}i\pi}(a_{-k}^* \cos\theta_k - a_{-k} \sin\theta_k) \qquad . \qquad (3.6.32b)$$

The angles θ_k are chosen so as to eliminate non-diagonal terms of the kind $a_k^* a_{-k}^*$ and $a_{-k}a_k$ in the Hamiltonian. The reader can verify that this will achieved if the choice:

$$\tan 2\theta_k = \frac{J \sin k}{(J \cos k - B)} \qquad (3.6.32c)$$

is made. He will then find that H_k is now in the form:

$$H_k = \Delta_k(a_k^* a_k + a_{-k}^* a_{-k} - 1) \quad , \qquad (3.6.33)$$

in total agreement with our earlier analysis: the vacuum has energy $-\Delta_k$, the 2-particle state $+\Delta_k$, and the two singly-occupied states have zero energy.

. . . .

Problem 3.10. Show that (3.6.32a) yields the transformation (3.6.32b) by letting op equal a_k, a_{-k}, a_k^* and a_{-k}^* in turn and computing the series in θ_k.

. . . .

The calculation of correlation functions is now straightforward. As a simple example, we compute $\langle S_j^+ S_{j+1}^- \rangle_{TA}$, which by (3.6.10) is the expectation value of the pair of fermions $c_j^* c_{j+1}$. We expand each in plane wave operators, then perform the Bogolubov transformation explicitly.

$$c_j^* c_{j+1} \rightarrow \frac{1}{N} \sum_{k,k'} \exp(-ikj) \exp[ik'(j+1)] \, a_k^* a_{k'}$$

$$\rightarrow \frac{1}{N} \sum_{k,k'} \exp(-ikj) \exp[ik'(j+1)](a_k^* \cos\theta_k + a_{-k} \sin\theta_k)$$

$$\times (a_{k'} \cos\theta_{k'} + a_{-k'}^* \sin\theta_{k'}) \qquad (3.6.34)$$

with the angles determined in (3.6.32). The thermal average is taken in the occupation-number representation, in which

$$\langle a_k^* a_{k'} \rangle_{TA} = \delta_{k,k'} (1 + e^{\beta\Delta_k})^{-1} \quad . \qquad (3.6.35)$$

After some algebra, we find

$$\langle S_j^+ S_{j+1}^- \rangle_{TA} = -\frac{1}{2\pi} \int_{-\pi}^{+\pi} dk \cos k (J \cos k - B) \frac{\tanh(\beta\Delta_k/2)}{\Delta_k} \quad . \qquad (3.6.36)$$

The generalization to non-nearest-neighbor spins can be made on the following pattern. $\langle S_0^+ S_p^- \rangle_{TA}$ is first written in terms of the fermions, by (3.6.7):

$$\langle S_0^+ S_p^- \rangle_{TA} = \left\langle c_0^* \exp\left(i\pi \sum_{0 \leqslant r < p} n_r\right) c_p \right\rangle_{TA}$$

$$= \langle c_0^*(c_1^* + c_1)(c_1^* - c_1) \cdots (c_{p-1}^* - c_{p-1}) c_p \rangle_{TA} \quad .$$

The c_j's are expanded in plane wave operators a_k which are then transformed by (3.6.32). The average over resultant multinomials in the a_k's is now taken, in the representation in which H is diagonal. Thus, only pairings of type (3.6.35) survive. An explicit example of this type calculation occurs in connection with the spontaneous magnetization in the 2D Ising model, and is detailed later in this chapter.

Let us now examine time-dependent correlations. Because the time-developed state $|t)$ satisfies Schrödinger's equation,

$$H|t) = \hbar i \frac{\partial}{\partial t} |t) \qquad (3.6.37)$$

it may be given the formal solution

$$|t) = e^{-iHt}|0) \qquad (3.6.38)$$

the only normalized state satisfying the initial condition. It follows that the probability of decaying into a state $|\alpha)$ at time t is:

$$P_\alpha(t) \equiv |(\alpha|e^{-iHt}|0)|^2 \quad . \qquad (3.6.39)$$

In particular, we are interested in the probability that the system stays in the initial state.

The probability of remaining in $|0)$ is:

$$P_0(t) = |(0|e^{-iHt}|0)|^2 \tag{3.6.40}$$

and the decay which occurs is measured by $1 - P_0(t)$. The H which appears above is given in (3.6.12). As we have seen, there is a sequence of unitary transformations which brings this H into diagonal form, denoted H_D; i.e., there exists an operator Ω such that, with $H(3.6.12)$ denoting H given by (3.6.12),

$$e^{-i\Omega}H(3.6.12)e^{i\Omega} = H_D \tag{3.6.41}$$

and therefore,

$$P_0(t) = |(0|e^{i\Omega}e^{-iH_Dt}\,e^{-i\Omega}\,e^{iH_Dt}\,e^{-iH_Dt}\,|0)|^2$$

$$= |(0|e^{i\Omega(0)}\,e^{-i\Omega(t)}\,e^{-iH_Dt}\,|0)|^2 \quad , \tag{3.6.42}$$

defining

$$\Omega(t) = e^{-iH_Dt}\,\Omega\,e^{+iH_Dt} \tag{3.6.43}$$

in a natural way. At this point, (3.6.32) is Ω, and we find

$$P_0(t) = \prod_{k>0} |(0|\exp[\theta_k(a_k^*a_{-k}^* - H.C.)]\,\exp[-\theta_k(a_k^*a_{-k}^*\,e^{-i\Delta_k t} - H.C.)]|0)|^2$$

$$= \exp\left[-\sum_{k>0} \ln(1 - \sin^2 2\theta_k\,\sin^2 \tfrac{1}{2}\Delta_k t)^{-1}\right]$$

$$= \exp\left[-\frac{N}{2\pi}\int_0^\pi dk\,\ln(1 - \sin^2 2\theta_k\,\sin^2 \tfrac{1}{2}\Delta_k t)^{-1}\right] \quad . \tag{3.6.44}$$

When $J = 0$, the state $|0)$ as defined above is the ground state, the parameters θ_k are all zero, and $P(t) \equiv 1$. When $J \neq 0$, the probability $P(t)$ becomes zero instantly [in time $O(1/N)$] because of the factor N in the exponent; there are no recursion times (Poincaré cycles) when the sum is replaced by an integral as in the equation above, as is proper in the thermodynamic limit.

If one studies states which are closer to the eigenstates of H, the decay should be less dramatic. The reader interested in magnetic relaxation will find a substantial literature [3.17].

We now turn to some purely mathematical considerations generalizing the solutions found here. The casual reader will wish to skip the next section, and turn instead to the theory of the two-dimensional Ising model which follows it.

3.7 Concerning Quadratic Forms of Fermion Operators

In the preceding, we ignored boundary conditions, variable J_n's and B_n's and yet other complications, in order to acquaint the reader with the salient features of what turned out to be, after all, a rather involved problem. This section is devoted to the "loose ends" and to some esoterica.

First, concerning boundary conditions, if we really have *free ends* we cannot use plane waves and the simplifying transformation (3.6.14) which permits the decoupling into normal modes, is inapplicable. If we have *periodic boundary conditions*, then the fermion Hamiltonian possesses a subtly more complicated structure than we had supposed, and there may exist *two* ground states. Let us re-examine the Jordan-Wigner transformation of (3.6.5), assuming the Nth spin is bonded to the first spin. We need to know operators such as $S_N^\pm S_1^\pm$ in Fermi operator language. By using the identities $\exp(i\pi n_N) c_N = c_N$ and $\exp(i\pi n_N) c_N^* = -c_N^*$ we can incorporate *all* the occupation number operators into the phase factors:

$$S_N^- S_1^+ \to e^{i\pi N} c_N c_1^* \quad \text{and}$$

$$\tag{3.7.1}$$

$$S_N^+ S_1^+ \to -e^{i\pi N} c_N^* c_1^* \quad ,$$

($S_N^- S_1^-$ and $S_N^+ S_1^-$ are obtainable by Hermitean conjugation). The operator N is the total occupation-number operator. Although it does not commute with the Hamiltonian [which creates and destroys particles in pairs, cf. (3.6.18,20)] parity *is* conserved (it is even or it is odd) H does not connect even occupation numbers to odd ones, and vice versa. Therefore the phase factors in (3.7.1) are both +1 for *even* occupancies and both -1 for *odd* occupancies.

In general, the two ground states will not be degenerate, but we do know that when B = 0 the Ising ground state is two-fold degenerate. It is this degeneracy which persists, in the thermodynamic limit of long chains (N → ∞), out to the critical point $|B/J| = 1$. For $|B| > |J|$, the ground state is unique, and increasingly oriented in the direction of B as this quantity is further increased.

It is important to note that although we *start* with a unique, given, spin Hamiltonian, it is transformed into *two* distinct quadratic forms in fermions. The one corresponding to even-occupancy states has a different operator representing the dynamics of the (N,1) bond than does the quadratic form representing the odd-occupancy states. *Both* differ from the typical bond (n,n+1). The even-occupancy fermion Hamiltonian should not be used to study odd-occupancy states, nor should the odd-occupancy fermion Hamiltonian be used in the calculation of even-occupancy states. If the total Hilbert space contains 2^N configurations, then each of the two distinct fermion Hamiltonians is applicable only in a subspace of $1/2 \times 2^N$ states. Certainly, the differences are small (1 bond out of N) and negligible in many cases, so that using only the periodic fermion Hamiltonian appropriate to the even-occupancy states, as in the preceding section, is adequate for many purposes.

But this entire situation is interesting in that it forces us to examine the case of fermion quadratic forms which are not explicitly homogeneous. In general, we may have arbitrary J_n's and B_n's, and we require a systematic method —not based on plane waves —to calculate the eigenvalue spectrum. The procedure we now outline is taken almost *verbatim* from [Ref.3.18, App. A].

We wish to diagonalize the quadratic form

$$H = \sum_{ij} [c_i^* A_{ij} c_j + \frac{1}{2} (c_i^* B_{ij} c_j^* + H.C.)] \tag{3.7.2}$$

where the c's and c^*'s are fermion anihilation and creation operators, A_{ij} is a real symmetric matrix and B_{ij} is a real antisymmetric matrix. (If they are not in this form, trivial phase transformations of the type $c_j, c_j^* \rightarrow \exp[i\phi(j)]c_j$, $\exp[-i\phi(j)]c_j^*$ can make them so if H is Hermitean.)

....

Problem 3.11. Verify that H in (3.6.13) is in the canonical form defined above in (3.7.2). Also, with k and -k replacing i and j, show how (3.6.20) may be put in canonical form by a phase transformation.

....

We try to find a linear transformation of the form,

$$b_k = \sum_i (g_{ki} c_i + h_{ki} c_i^*) \quad \text{and}$$

$$b_k^* = \sum_i (g_{ki} c_i^* + h_{ki} c_i) \quad , \tag{3.7.3}$$

with real g_{ki} and h_{ki} coefficients, which is canonical so that the b_k's and b_k^*'s satisfy the fermion anticommutation relations (3.6.15), and which diagonalizes H:

$$H = \sum_k E_k b_k^* b_k + \text{const.} \tag{3.7.4}$$

If this is possible, b_k is a *lowering operator* of H and satisfies:

$$[b_k, H] = E_k b_k \quad . \tag{3.7.5}$$

Substituting (3.7.3) into (3.7.5) and equating coefficients of each operator c_i and c_i^* we obtain sets of equations for the g's and h's:

$$\sum_j (g_{kj} A_{ji} - h_{kj} B_{ji}) = E_k g_{ki} \quad \text{and}$$

$$\sum_j (g_{kj} B_{ji} - h_{kj} A_{ji}) = E_k h_{ki} \quad . \tag{3.7.6}$$

These are simplified by means of the linear combinations

$$\phi_{ki} = g_{ki} + h_{ki} \quad \text{and} \quad \psi_{ki} = g_{ki} - h_{ki} \tag{3.7.7}$$

in terms of which the coupled equations are:

$$\phi_k(A - B) = E_k \psi_k \quad \text{and}$$

$$\psi_k(A + B) = E_k \phi_k \tag{3.7.8}$$

in obvious matrix notation. Eliminating ϕ or ψ, we have either

$$\phi_k(A - B)(A + B) = E_k^2 \phi_k \quad \text{or}$$

$$\psi_k(A + B)(A - B) = E_k^2 \psi_k \quad . \tag{3.7.9}$$

For $E_k = 0$, (3.7.8) provides the solutions. For $E_k \neq 0$, one may solve the eigenvalue equations (3.7.9) and use (3.7.8) to relate the phase of the two solutions.

Because A is symmetric and B antisymmetric, $(A + B)^T = A - B$ so that the product matrices in (3.7.9) are symmetric and non-negative. With all the eigenvalues E_k real, one can choose the eigenvectors to be real as well as orthogonal. If the ϕ_k's are normalized ($\sum_i \phi_{ki}^2 = 1$) then the ψ_k's are also, or can be chosen to be so. This ensures that

$$\sum_i (g_{ki} g_{k'i} + h_{ki} h_{k'i}) = \delta_{kk'} \tag{3.7.9a}$$

and

$$\sum_i (g_{ki} h_{k'i} - g_{k'i} h_{ki}) = 0 \quad , \tag{3.7.9b}$$

the necessary and sufficient condition for the b_k, b_k^* to satisfy the fermion anticommutation relations.

· The energies E_k are thus the solutions of the equations above. The constant in H, (3.7.4), is obtained by the invariance of the trace of H, and thus we have for (3.7.4):

$$H = \sum_k E_k b_k^* b_k + \frac{1}{2} \left(\sum_i A_{ii} - \sum_k E_k \right) . \tag{3.7.10}$$

Now consider the case of arbitrary bonds connecting nearest-neighbor spins J_n and arbitrary fields B_n, as recently solved by Pfeuty [3.19]. The **A - B** matrix takes the form

$$\mathbf{A} - \mathbf{B} = \begin{bmatrix} B_1 & \cdot & \cdot & J_N \\ J_1 & B_2 & 0 & \vdots \\ & J_2 & \cdot & \\ 0 & & \cdot & \cdot \\ & & \cdot & B_N \end{bmatrix} . \tag{3.7.11}$$

All the B_n's and J_n's can be made ≥ 0 by appropriate rotations of each of the spins through 180° where necessary.

The secular equations (3.7.9) must, in general, be solved numerically. We may, however, inquire when is there a switch-over from J_n to B_n domination, i.e. where is the critical point analogous to $|J/B| = 1$ of the homogeneous case. Evidently, this occurs when an eigenvalue is zero, i.e. when the determinant of the product matrix vanishes:

$$\text{Det}\| (\mathbf{A} - \mathbf{B})(\mathbf{A} + \mathbf{B})\| = 0$$

but because **A - B** is the transpose of **A + B** it suffices to determine the point at which

$$\text{Det}\|\mathbf{A} - \mathbf{B}\| = \prod_n |B_n| - \prod_n |J_n| = 0 \tag{3.7.12}$$

vanishes. Taking logarithms, we establish the *critical point* for *any* given ensemble of *arbitrary or random* B_n's and J_n's *as*

$$<\ln|B_n|> = <\ln|J_n|> , \tag{3.7.13}$$

where the averages are taken over the given ensemble.

After diagonalization, one often wishes to calculate ground-state or thermal-averaged correlation functions. This can be done directly with the matrices introduced in this section, but the algebra becomes rather involved and the interested reader is referred to [3.18].

Finally, from the title of this section it is clear that we can do nothing for next-nearest neighbor interactions (quartic in fermions), spins greater than spins 1/2 (fermions are altogether inappropriate there), or the full Heisenberg model. However, two generalizations of the Ising model: the

XY and the Heisenberg-Ising chain *can* be reduced to quadratic forms in fermions, and were solved in [3.18].

3.8 Two-Dimensional Ising Model: The Transfer Matrix

The manner in which one builds up a square lattice starting from linear chains dictates the construction of the transfer matrix in the two-dimensional Ising model. Figure 3.7 illustrates the process of laying out the chains horizontally, one below the next. The bonds *within* the jth chain are $J^x_{j,1}$, $J^x_{j,2}$,..., $J^x_{j,n}$,...$J^x_{j,N-1}$ with the superscript x referring to the horizontal direction of the bond, the first subscript to the chain and the second subscript to the two spins $(n,n+1)$ which are connected by it. (In treating the single chain of Sect.3.5 we could dispense with both superscript x and chain index j, and wrote such bonds merely as J_n.)

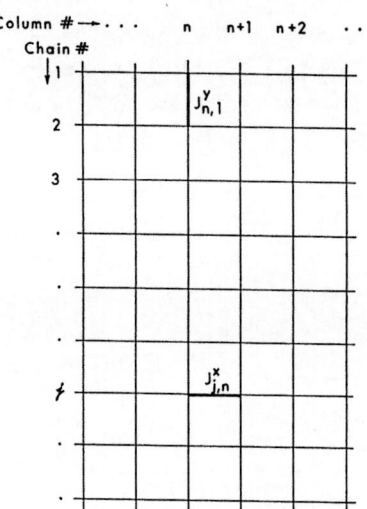

Column # →··· n n+1 n+2 ···
Chain #
↓1
2
3
·
·
·
≠
·
·

$J^y_{n,1}$

$J^x_{j,n}$

Fig.3.7. Arbitrary bonds J^x (horizontal) and J^y (vertical) on sq lattice. In the homogeneous case, these are labeled J_1 and J_2

The feature which distinguishes the two-dimensional model from an array of independent chains is the presence of vertical bonds, which we label $J^y_{n,j}$. Here y indicates the direction of the bond (now vertical), n the column, and j the pair of spins in the jth and j+1-st chains which are connected by it. In addition to the horizontal bonds, the transfer matrix must incorporate these inter-chain connections.

Let us start with the horizontal contributions. The transfer matrix of (3.5.9) V_n must now be relabeled, $V^x_{j,n}$ and the total transfer operator is the product over all chains,

$$\prod_{j=1}^{N} \mathbf{v}^x_{j,n} \equiv \mathbf{v}^x_n \quad . \tag{3.8.1}$$

The vertical contributions, i.e. the bonds which are included in the nth column, carry a Boltzmann factor which we denote $\mathbf{v}^y_{n,j}$, each being

$$\mathbf{v}^y_{n,j} \equiv \exp(\beta J^y_{n,j} S^z_{n,j} S^z_{n,j+1})$$

and the total being

$$\mathbf{v}^y_n = \prod_{j=1}^{N} \mathbf{v}^y_{n,j} \quad . \tag{3.8.2}$$

Defining the transfer matrix as the combination

$$\mathbf{V}_n \equiv \mathbf{v}^x_n \mathbf{v}^y_n \quad , \tag{3.8.3}$$

we can evaluate the partition function

$$Z = \mathrm{Tr}\{\mathbf{V}_1 \mathbf{V}_2 \cdots \mathbf{V}_{N-1}\} \tag{3.8.4}$$

for an array of N columns. If the last column is connected to the first (periodic boundary conditions) we can label the additional transfer operator $\mathbf{V}_N$, and include it also.

Of course, it would be desirable to be able to evaluate Z for arbitrary $J^x_{n,j}$ and $J^y_{n,j}$, but this is intractable. The most general problem we shall face has all J^x bonds equal, denoted J_1 henceforth, and all vertical (J^y) bonds equal to a value J_2, possibly different from J_1, the *anisotropic* square lattice. With the magnetic field B_n also constant, the trace in (3.8.4) can be evaluated as λ^N, where λ is the optimal solution of the eigenvalue problem,

$$\mathbf{V} \cdot \mathbf{u} = \lambda \mathbf{u} \quad . \tag{3.8.5}$$

$\mathbf{V}$ is a matrix in 2^N dimensions, λ is its largest eigenvalue, and $\mathbf{u}$ the corresponding eigenvector.

Suppose the lattice is rectangular, N_x wide by N_y high. The eigenvalue λ must be of the form,

$$\lambda = \exp(AN_y)$$

and the partition function will be

$$Z = [\exp(AN_y)]N_x = \exp(AN_x N_y)$$

demonstrating that the free energy is properly extensive, $F = f N_x N_y$. Indeed, A is just $-f/kT$. Once again the requirement of largest λ is equivalent to optimization of the free energy, a postulate of statistical mechanics stressed in Chap.2.

Although at this point the eigenvalue problem is well posed and its solution will yield Z, hence f and the various thermodynamic functions that are calculated as derivatives of f, the form of V^x leaves something to be desired. This operator includes B and is complicated even in homogeneous fields (B_n = const.). It so happens that we can simplify the problem considerably by including the external field terms with the vertical bonds, in V^y. We therefore reformulate, starting with H. Let

$$H = \sum_n H_n \tag{3.8.6}$$

where each H_n contains all the new bonds introduced when the n+1-st column is added, and includes the interactions of spins in that column with any external field. Thus, $H_n = H_n^x + H_n^y + H_n^B$, where the horizontal bonds are contained in H_n^x,

$$H_n^x = -J_1 \sum_{j=1}^{N} S_{n,j}^z S_{n+1,j}^z \quad , \tag{3.8.7a}$$

the vertical bonds being

$$H_n^y = -J_2 \sum_{j=1}^{N-1} S_{n+1,j}^z S_{n+1,j+1}^z \quad . \tag{3.8.7b}$$

Finally, the interactions which the spins in the new column have with an external field are included in

$$H_n^B = -B \sum_{j=1}^{N} S_{n+1,j}^z \quad . \tag{3.8.7c}$$

The reader will verify that a sum of the above operators over all n yields the total Hamiltonian for the model, with the possible exception of boundary effects which are irrelevant to the computation of f, the free energy per spin.

Exponentiating the H^x operator, we can use (3.5.9) with B = 0 to obtain, first for the individual chains,

$$V_{j,n}^x = \exp(\beta J_1)\mathbf{1} + \exp(-\beta J_1)\sigma_j^x = [2 \sinh(2\beta J_1)]^{\frac{1}{2}} \exp(K_1^*\sigma_j^x) \quad . \tag{3.8.8}$$

This serves to define the real parameter K_1^*, using the identity $\exp(K^*\sigma^x)$ $= \cosh K^* + \sigma^x \sinh K^*$, and the results of Problem 3.12.

....

Problem 3.12. Verify that (3.8.8) requires that K_1^* satisfies:

$$\tanh K_1^* = \exp(-2K_1) \quad \text{(with } K_1 \equiv \beta J_1\text{)} \quad , \tag{3.8.9}$$

and also check the relations,

$$\tanh K_1 = \exp(-2K_1^*) \quad \text{and} \quad \sinh 2K_1 \sinh 2K_1^* = 1 \tag{3.8.10}$$

which are precisely the *duality* relations of Sect.3.3.
....

With V_j^X in the form of the rhs of (3.8.8) it is an easy matter to include the contributions for all chains $j = 1, 2, \ldots$:

$$V^X = [2 \sinh(2K_1)]^{\frac{1}{2}N} \exp\left(K_1^* \sum_j \sigma_j^X\right) \quad . \tag{3.8.11a}$$

The interchain contributions are summed in

$$V^y = \exp\left(K_2 \sum_j \sigma_j^z \sigma_{j+1}^z\right) , \tag{3.8.11b}$$

and the external field appears in

$$V^B = \exp\left(h \sum_j \sigma_j^z\right) \tag{3.8.11c}$$

with $K_2 = \beta J_2$ and $h = \beta B$, in an obvious notation, and $\sigma_j^z = S_j^z$.

Now, the eigenvalue equation (3.8.5) with $V = V^X V^y V^B$ is simple, except for minor inconveniences in dealing with the exponentials (3.8.11a) of operators which do not commute with other exponentials (3.8.11b or c).

If one just ignored the lack of commutation and combined the three exponents, the combined term takes on the appearance of the 1D Ising model in *both* transverse and longitudinal fields (K_1^* being the effective transverse field, and h the longitudinal). In Sect.3.5 we analyzed the latter, in Sect. 3.6 the former. The difficulties of obtaining the eigenstates analytically in the presence of both fields was apparent. For this reason, it is convenient to proceed with $B = 0$, obtaining the magnetic properties in terms of zero-field correlation functions. In finite B, the appropriate analysis will be numerical. In Sect.3.11, in which *complex* fields are analyzed, the transfer matrix formalism is abandoned altogether.

In selecting the form of V to use in the eigenvalue equation (3.8.5), we have several options which differ from each other by similarity transformations. The natural choice, $V^X V^y V^B$ is not self-adjoint. This is an inconvenience if we seek to calculate correlation functions, as u^T will not be the left-hand eigenvector. There *are* a number of self-adjoint combinations which lead to the same optimum eigenvalue, and

$$(V^y V^B)^{\frac{1}{2}} V^x (V^y V^B)^{\frac{1}{2}} \quad \text{and} \quad V^{x\frac{1}{2}} (V^y V^B) V^{x\frac{1}{2}} \tag{3.8.12}$$

are just two of them. Because V^y and V^B involve the same Pauli matrices, they commute and can be written in any order, or combined into a single exponential form.

Although the derivation of the transfer matrix and the resulting discussions have centered about the 2D case, the generalization to 3D or higher, or to non-nearest neighbor bonds is obvious. It suffices to include in V^y *all* the bonds which go into growing the lattice layer by layer. In 3D, the column becomes a plane; each spin must be labeled by two indices, its position in the plane. In 4D it is a volume and 3 scripts are required. In d dimensions, the transfer matrix describes a d-1-dimensional Ising model subject to the transverse operators in V^x; not surprisingly, these are called "disorder operators", and are most important at high temperature (K^* increases with T, while K decreases). The operators in V^y and V^B are the "order operators", dominating at low temperature.

Where it is solvable, the transfer matrix formalism simplifies the statistical mechanics. We need only the largest eigenvalue of a model in one fewer dimension, complicated only by the presence of the new disorder operators.

There exists an alternative derivation of the transfer matrix formalism, somewhat more natural than the preceding as it clarifies the meaning of the eigenvector **u** in (3.8.5), the eigenvector belonging to the optimum eigenvalue λ, henceforth denoted the optimal eigenvector.

Frobenius' theorem concerning the optimum eigenvector of a matrix, all elements of which are positive, states that all the components of the optimum eigenvector are themselves positive. The proof is intuitive: any change in sign can only lower the eigenvalue.

It follows that we can consider the various elements u_j of **u** as probabilities, provided the eigenvector is normalized according to the rule,

$$\sum_j u_j = 1 \quad . \tag{3.8.13}$$

It is also apparent that any vector with all positive elements is not orthogonal to **u**. Suppose we start the first column in a 2D Ising model in total disorder, corresponding to each element in $\mathbf{u}_1$ being $u_{1j} = 2^{-N}$. With this choice, all configurations appear with equal probability. If we expand $\mathbf{u}_1$ in eigenvectors of **V**, the optimum eigenvector **u** appears with a coefficient a_0:

$$\mathbf{u}_1 = a_0\mathbf{u} + a_1\mathbf{u}' + a_2\mathbf{u}'' + \ldots \tag{3.8.14}$$

indicating the other eigenvectors by $\mathbf{u}'$, $\mathbf{u}''$, etc. Applying the transfer matrix n times brings us to the n+1-st column, and yields

$$\mathbf{v}^n \cdot \mathbf{u}_1 = a_0\lambda^n\mathbf{u} + a_1\lambda'^n\mathbf{u}' + a_2\lambda''^n\mathbf{u}'' + \ldots \tag{3.8.15}$$

Regardless of the original values of a_1, a_2, etc., the resulting coefficients $a_1\lambda'^n$, $a_2\lambda''^n$, etc. all tend exponentially to zero by comparison with the optimum value $a_0\lambda^n$, which alone survives in the asymptotic limit $n \to \infty$. (The possibility $a_0 = 0$ is eliminated, by noting that two vectors $\mathbf{u}$ and $\mathbf{u}_1$, both of which have all positive elements, cannot be orthogonal.)

Thus, even if we started with disorder, $u_{1j} = 2^{-N}$, repeated application of the transfer matrix brings the nth vector into the form of the optimum eigenvector, reflecting the probabilities of having the various configurations at the given temperature T. The optimal eigenvector is the "reduced density matrix" for the problem.

Repeated applications of V are tantamount to the approach to thermal equilibrium. This approach is fastest when the leading eigenvalue is separated from the others by a finite gap: $\lambda'/\lambda < 1$. This is generally the case, and convergence by this method is far more efficient than the usual solution of a secular determinant. The exception: at T_c, in the absence of any applied field, the gap disappears and the approach to thermal equilibrium is critically damped. With this exception, the iterative approach to the largest eigenvalue is the method of choice when numerical calculations are required. Of course, it is not required to start with a perfectly random configuration. Once $\mathbf{u}$ is known for a given set of parameters (such as T or B) *it* can serve as the initial $\mathbf{u}_1$ for different parameters (T + dT, B + dB), and convergence is enhanced.

3.9 Solution of Two-Dimensional Ising Model in Zero Field

It is particularly convenient to rotate the spin operators, so that $\sigma^x \to -\sigma^z$ and $\sigma^z \to +\sigma^x$ just as in (3.6.12), and obtain

$$\mathbf{v}^x = [2 \sinh(2K_1)]^{\frac{1}{2}N} \exp\left(-K_1^* \sum_j \sigma_j^z\right) , \tag{3.9.1a}$$

$$\mathbf{v}^y = \exp\left(K_2 \sum_j \sigma_j^x \sigma_{j+1}^x\right) \tag{3.9.1b}$$

and

$$\mathbf{V}^B = \exp\!\left(h \sum_j \sigma_j^x\right) \; . \tag{3.9.1c}$$

In zero field, $h = 0$, and $\mathbf{V}^B = 1$. The exponents in (3.9.1a and b) are precisely the two terms in (3.6.12) — with K_1^* replacing B in the earlier equation, and K_2 replacing J. Our method of solution will be similar in its outline [3.20].

First, writing $\sigma_j^x = S_j^+ + S_j^-$ and $\sigma_j^z = 2S_j^+ S_j^- - 1$, then invoking the Jordan-Wigner transformation to fermion operators c_j and c_j^*, we obtain:

$$\mathbf{V}^x = [2 \sinh(2K_1)]^{\frac{1}{2}N} \exp\!\left[-2K_1^* \sum_j \left(c_j^* c_j - \tfrac{1}{2}\right)\right] \tag{3.9.2a}$$

and

$$\mathbf{V}^y = \exp\!\left[K_2 \sum_j (c_j^* - c_j)(c_{j+1}^* + c_{j+1})\right] \; . \tag{3.9.2b}$$

In Sect.3.7 we already noted that the boundary conditions may be affected by the Jordan-Wigner transformation. Supposing that we want the *spin* Hamiltonian to be periodic in the y direction (forming a cylinder of circumference N and length N) then mathematically we can retain the form of the exponentials in fermions in (3.9.2) without modification, the bond connecting $N \rightarrow N+1$ being the same as the others, provided we identify c_{N+1} with c_1 in the correct way. This requires:

$$c_{N+1} = -c_1 \quad \text{and} \quad c_{N+1}^* = -c_1^*$$

for even occupancy states, $\tag{3.9.3a}$

whereas

$$c_{N+1} = c_1 \quad \text{and} \quad c_{N+1}^* = c_1^*$$

for odd-occupancy states . $\tag{3.9.3b}$

Both are achievable by an expansion in plane waves; the first by the use of anti-periodic boundary conditions, the second with the more conventional periodic boundary conditions. Let us label the transfer matrix for the even occupancy case $\mathbf{V}^+$, for the odd-occupancy case, $\mathbf{V}^-$. The odd eigenstates of $\mathbf{V}^+$ should be discarded, and likewise, the even eigenstates of $\mathbf{V}^-$ are also to be discarded.

The second possible boundary condition on the spins, free ends, is conceptually simpler: the N-to-1 bond is absent; the natural waveforms are sin kj rather than the plane waves. We shall leave the analysis of this case (which should parallel closely the following derivations for the periodic/antiperiodic cases) as an exercise for the reader, with Sect.3.7 as a guide.

Because of translational symmetry, we again try a plane-wave expansion, writing the c_j's in the form

$$c_j = \frac{1}{N^{\frac{1}{2}}} e^{-\frac{1}{4}i\pi} \sum_q e^{iqj} a_q \qquad (3.9.4)$$

and similarly for the c_j^*'s. The factor $\exp(-i\pi/4)$ is incorporated in anticipation of a transformation similar to (3.6.32). The a_q's are a set of fermion operators, with values of q chosen to satisfy anticyclic boundary conditions for even occupancy, or cyclic boundary conditions for odd occupancy. That is, from (3.9.3a) above, we see $\exp(iqN) = -1$, and

$$q = \pm\pi/N \ , \ \pm3\pi/N \ ,\ldots, \ \pm(N - 1)\pi/N \ , \qquad (3.9.5a)$$

whereas from (3.9.3b) with $\exp(iqN) = +1$, we obtain the more conventional

$$q = 0 \ , \ \pm2\pi/N \ , \ \pm4\pi/N \ ,\ldots, \ \pm(N - 2)\pi/N \ , \ +\pi \qquad . \qquad (3.9.5b)$$

Each of V^x and V^y is now of the form $\prod V_q^x$ and $\prod V_q^y$ with $0 \leqslant q \leqslant \pi$. Factoring out the ubiquitous $[2\sinh(2K_1)]^{\frac{1}{2}N}$, we have simply

$$V_q^x = \exp[-2K_1^*(a_q^* a_q + a_{-q}^* a_{-q} - 1)] \quad \text{and} \qquad (3.9.6)$$

$$V_q^y = \exp\{2K_2[\cos q(a_q^* a_q + a_{-q}^* a_{-1} - 1) + \sin q(a_q a_{-q} + a_{-q}^* a_q^*)]\} \qquad (3.9.7)$$

for $q \neq 0$ or π. (In those special cases, for which there is not -q, we have

$$V_0 = \exp\left[-2(K_1^* - K_2)\left(a_0^* a_0 - \frac{1}{2}\right)\right] \ , \quad V_\pi = \exp\left[-2(K_1^* + K_2)\left(a_\pi^* a_\pi - \frac{1}{2}\right)\right] \ , \qquad (3.9.8)$$

combining V^x and V^y at these points where, exceptionally, they commute.)

It is also possible to combine V_q^x and V_q^y at arbitrary q by some sort of Baker-Hausdorff expansion. Many algorithms have been devised to combine exponential operators into a single exponential or, conversely, to disentangle an exponential of noncommuting operators into products of exponentials. One such relation, the Zassenhaus formula, is

$$\exp[g(A + B)] = \exp(gA) \exp(gB) \exp(g^2 C_2) \exp(g^3 C_3) \ldots \exp(g^n C_n) \ldots \qquad (3.9.9)$$

where

$$C_2 = \frac{1}{2} [B,A] \ , \quad C_3 = \frac{1}{6} [C_2,A + 2B] \ , \ \ldots$$

$$C_n = \frac{1}{n!} \left[\frac{d^n}{dg^n} \left(\exp(-g^{n-1}C_{n-1}) \ldots \exp(-g^2 C_2)\right.\right.$$

$$\left.\left. \times \ \exp(-gB) \exp(-gA) \exp[g(A + B)]\right)\right]_{g=0} \ . \qquad (3.9.10)$$

Read from left to right, (3.9.9) is used to disentangle A and B; from right to left, to combine them. This formula has many uses beyond its present application.

For our purposes, it is sufficient to note that if A and B are bilinear in fermion operators, so are C_2, C_3, and all the other C_n's. In fact, they are *all* linear combinations of what amounts to Pauli spin matrices in occupation-number space, as explicitly demonstrated in Problem 3.13.

....

Problem 3.13. With A and B any linear combinations of the operators:

$$\underset{\sim}{X} \equiv a^*_{-q}a^*_q + a_q a_{-q} \ , \quad \underset{\sim}{Y} \equiv -i(a^*_{-q}a^*_q - a_q a_{-q}) \ , \quad \underset{\sim}{Z} \equiv n_q + n_{-q} - 1$$

show that C_2, C_3,... will all be of this form also. Show that these three operators are the Pauli matrices in the space $|0\rangle$, $a^*_{-q}a^*_q|0\rangle$, but all *vanish* in the space of $a^*_q|0\rangle$ and $a^*_{-q}|0\rangle$.

....

Thus, with V^x_q of the form $\exp(-2K_1\underset{\sim}{Z})$ and $V^y_q = \exp\{2K_2[(\cos q)\underset{\sim}{Z} + (\sin q)\underset{\sim}{X}]\}$, it is clear that we can combine them into a single exponential of the form

$$V^{y\frac{1}{2}}_q V^x_q V^{y\frac{1}{2}}_q = c \ \exp(a\underset{\sim}{X} + b\underset{\sim}{Z}) \tag{3.9.11}$$

(a,b,c are constants) which can then be brought into diagonal form by a spin rotation. Since the lhs of this equation yields 1 for the odd-occupancy states, as so does the rhs, this equality would extend to *all* four states in the q,-q subspace. Rather than use the Pauli spin algebra to resolve our difficulties, we can use an elegant trick. Once V is in diagonal form, it must be

$$\mathbf{V}_q = c \ \exp[\varepsilon_q(n_q + n_{-q} - 1)] \tag{3.9.12}$$

where c and ε_q are to be determined. With $\mathbf{n}_q = b^*_q b_q$ and $\mathbf{n}_{-q} = b^*_{-q}b_{-q}$ and b_q a linear combination of a_q and a^*_{-q}, we know that

$$\mathbf{V}^{-1}_q b_q \mathbf{V}_q = b_q \ \exp(\varepsilon_q) \tag{3.9.13}$$

and similarly for b_{-q}. On the other hand, we can compute the same relation, with $b_q = a_q \cos\theta_q + a^*_{-q} \sin\theta_q$ and V_q in the form (3.9.6)^x(3.9.7). This yields an eigenvalue equation, the solution of which gives both θ_q and ε_q. Implementation is left to the diligent reader.

The conclusion is [3.20]:

$$\tan(2\theta_q) = \frac{2C(q)}{B(q) - A(q)} \ , \tag{3.9.14}$$

in which $\text{sgn}(2\theta_q) = \text{sgn}(q)$, and

$$A(q) = \exp(-2K_1^*)(\cosh K_2 + \sinh K_2 \cos q)^2 + \exp(2K_1^*)(\sinh K_2 \sin q)^2$$

(3.9.15a)

$$B(q) = \exp(-2K_1^*)(\sinh K_2 \sin q)^2 + \exp(2K_1^*)(\cosh K_2 - \sinh K_2 \cos q)^2$$

(3.9.15b)

$$C(q) = (2 \sinh K_2 \sin q) \times (\cosh 2K_1^* \cosh K_2 - \sinh 2K_1^* \sinh K_2 \cos q) \quad .$$

(3.9.15c)

Finally, ε_q is the positive root of

$$\cosh\varepsilon_q = \cosh 2K_2 \cosh 2K_1^* - \sinh 2K_2 \sinh 2K_1^* \cos q \quad . \qquad (3.9.16)$$

In this diagonal representation,

$$\mathbf{V} = (2 \sinh 2K_1)^N \exp\left[-\sum_{\text{all } q} |\varepsilon_q|(b_q^* b_q - \tfrac{1}{2})\right] \quad . \qquad (3.9.17)$$

For *even* total occupancy, the set (3.9.5a) of q's is to be used in the sum; for *odd* total occupancy, the set (b). Precisely at $K_1^* = K_2$ the eigenvalue of the q = 0 mode vanishes, so *it can be occupied or not*, without changing Z. At this temperature - and, below it as well, the vacua for even occupancy and for odd occupancy are degenerate in the large N limit. It is a simple algebraic exercise to verify: (i) only at this critical point $K_1^* = K_2$ is the spectrum gapless and linear at small q($\lim q \to 0 \; \varepsilon_q \sim |q|$), (ii) at all other temperatures there is an energy gap against excitations (of magnitude $|K_1^* - K_2|$), and (iii) that this yields the same critical temperature as the more familiar condition

$$\sinh(2J_1/kT_c) \sinh(2J_2/kT_c) = 1 \quad , \qquad (3.9.18)$$

i.e., $kT_c/J = 2.269185...$ for $J_1 = J_2 = J$. It is almost self-evident that Z is given by setting all $b_k^* b_k = 0$,

$$Z = (2 \sinh 2K_1)^N \exp\left(\tfrac{1}{2} \sum_q |\varepsilon_q|\right) \qquad (3.9.19)$$

and the free energy per spin, f, by

$$f = -kT\left[\ln(2 \sinh 2K_1)^{\frac{1}{2}} + \frac{1}{4\pi} \int_{-\pi}^{+\pi} dq |\varepsilon_q|\right] \quad . \qquad (3.9.20)$$

It is possible to solve for ε_q ((3.9.16) is for $\cosh\varepsilon_q$) and evaluate f. Although this must be done numerically, the derivatives of f can be expressed in terms of elliptic functions, and one finds that the second derivative — the specific heat — involves K [given in (3.6.29) and (3.6.31)] and therefore diverges logarithmically at T_c. But (3.9.20) is not manifestly symmetric in x and y, as it should be. A simple integral identity transforms it into a two-dimensional integral, fully symmetric in J_1 and J_2.

With ε_q replacing x in the identity (2.9.6) we have for the integral above

$$\frac{1}{4\pi} \int_{-\pi}^{+\pi} dq \, \frac{1}{2\pi} \int_{-\pi}^{+\pi} dq' \, \ln(2\cosh\varepsilon_q - 2\cos q')$$

into which we may insert the solution (3.9.16) for $\cosh\varepsilon_q$,

$$\frac{1}{4\pi} \int dq \, \frac{1}{2\pi} \int dq' \, \ln(2\cosh 2K_2 \cosh 2K_1^* - 2\sinh 2K_2 \sinh 2K_1^* \cos q - 2\cos q').$$

Factoring out $2\sinh 2K_1^* = (\frac{1}{2}\sinh 2K_1)^{-1}$, cf. (3.8.10), yields

$$-\frac{(2\pi)^2}{8\pi^2} \ln(\tfrac{1}{2}\sinh 2K_1) + \frac{1}{8\pi^2} \int dq_1 \int dq_2 \, \ln(\cosh 2K_2 \coth 2K_1^*$$

$$- \sinh 2K_1 \cos q_1 - \sinh 2K_2 \cos q_2)$$

so that finally, with the aid of the duality relation (3.8.9) and standard trigonometric identities we eliminate $\coth 2K_1^*$

$$\coth 2K_1^* = \frac{1 + \exp(-4K_1^*)}{1 - \exp(-4K_1^*)} = \frac{1 + \tanh^2 K_1}{1 - \tanh^2 K_1} = \cosh 2K_1 \quad,$$

to arrive at Onsager's expression:

$$f = -kT\left[\ln 2 + \frac{1}{2\pi^2} \int\!\!\int_0^\pi dq_1 dq_2 \, \ln(\cosh 2K_1 \cosh 2K_2 - \sinh 2K_1 \cos q_1\right.$$

$$\left. - \sinh 2K_2 \cos q_2)\right] \quad. \tag{3.9.21}$$

Despite a surprising resemblance to the 2-dimensional Gaussian model, (2.9.5b), this free energy is well-behaved both above and below T_c. A new feature — the hyperbolic functions, trade-marks of the Ising model — makes for better-behaved thermodynamic functions.

To compare with the Gaussian result, consider the isotropic case $J_1 = J_2$ for which T_c is given by $\sinh(2J/kT_c) = 1$.

If we write

$$\sinh 2K = 1 - t \quad, \tag{3.9.22}$$

we may consider t a small parameter in the critical region, and f can now be written

$$f_I \cong -kT\left\{\ln[2(1 - t)^{\frac{1}{2}}] + \frac{1}{2\pi^2} \int\!\!\int_0^\pi dq_1\, dq_2 \ln\left(\frac{t^2}{1 - t} + 2 - \cos q_1 - \cos q_2\right)\right\}$$

$$(3.9.23)$$

By contrast, the Gaussian model, (2.9.5b) yields

$$f_G \cong +\frac{1}{2}\, kT\left[\ln(2J/kT) + \frac{1}{2\pi^2} \int\!\!\int_0^\pi dq_1\, dq_2 \ln(\hat{t} + 2 - \cos q_1 - \cos q_2)\right]$$

with $\hat{t} = \frac{k}{2J}(T - T_G)$, where $T_G = 4J/k$ in 2D . $(3.9.24)$

(The spherical model can be written in similar form, with $\hat{t}$ involving μ (i.e., τ) as well, but as $T_{csph} = 0$ in 2D it is not interesting to pursue the analogy.)

So the main difference between the Ising and the Gaussian free energies is the overall ±sign. Secondarily, the singularity at t (or $\hat{t}$) = 0 is approached as t^2 in the former, and linearly as $\hat{t}$ in the latter. So while the internal energy diverges (unphysically!) in the Gaussian approximation, it is the specific heat —a second derivative of f with respect to t —which is weakly, logarithmically, singular at T_c in the present case.

The specific heat for this case and in the anisotropic $(J_1 \neq J_2)$ limits is sketched in Fig.3.8.

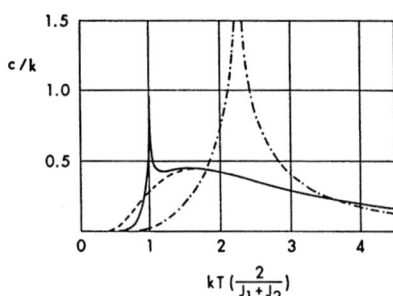

Fig.3.8. Specific heat for anisotropic 2D Ising model on sq lattice: $J_2 = J_1/100$ (——), limit of independent chains $J_2 = 0$ (----), as compared to the isotropic result, (3.9.26) (-·-··-·-) [3.9]

The exact formulas for the internal energy and specific heat of the Ising ferromagnet on an isotropic sq lattice are:

$$U/N = -J\, \coth(2K)\left[1 + \frac{2}{\pi}\, k_1'' K(k_1)\right] \quad \text{with} \tag{3.9.25}$$

$$k_1 = 2\, \sinh(2K)/\cosh^2 2K \quad \text{and}$$

$$k_1'' = 2\, \tanh^2 2K - 1 \quad .$$

135

The specific heat is logarithmically singular at T_c,

$$c/k_B = \frac{2}{\pi} (K \coth K)^2 \left\{ 2K(k_1) - 2E(k_1) - (1 - k_1'') \left[\frac{1}{2} \pi + k_1'' K(k_1) \right] \right\} \quad (3.9.26)$$

with the elliptic integrals given in (3.6.29).

It is not surprising that the same integrals occur in this study as in the ground-state properties of the one-dimensional Ising model in a transverse field. For if in the present section we had merely combined the exponentials in (3.9.1a,b) without regard to the commutators (3.9.9) which go into the exact expression, our problem would have reduced to that of Sect. 3.6. Indeed, the correspondences are:

Zero-Field 2D Ising Model ⟷ *1D Ising in Transverse Field*

$-f/kT$	E_0/N
K_1^*	B
K_2	J

The internal energy as calculated with the method of Sect.3.6 is compared with the exact results, (3.9.25), in Fig.3.9. The agreement is excellent at low T, at high T (and at T_c which is the same in the approximate and exact analyses).

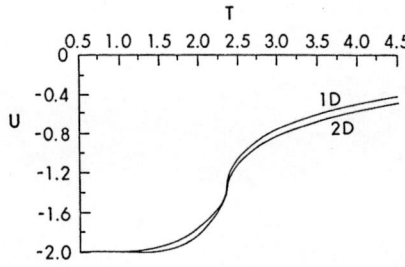

Fig.3.9. Plot of internal energy U(T) of isotropic Ising model on 2D sq lattice vs equivalent quantity obtained from ground state of 1D Ising model in transverse magnetic field. The agreement is exact at the three critical points: $T = 0$, T_c and ∞ [3.26]

In Table 3.2 below, we list some of the calculated properties of the Ising ferromagnet on three of the important two-dimensional lattices: sq, triangular and honeycomb (hexagonal). Although *anti*ferromagnets can also be studied by these same methods, their solutions yield surprises as we shall see subsequently.

In the following section, we turn to the magnetic properties: spontaneous magnetization of the ferromagnet below T_c, and magnetic susceptibility above T_c.

	Sq	Triangular	Honeycomb (Hex.)
$\exp(-2J/kT_c)$[a]	$2^{\frac{1}{2}} - 1$	$3^{-\frac{1}{2}}$	$2 - 3^{\frac{1}{2}}$
kT_c/J	2.269185	3.640957	1.518652
U_c/NJ (Internal energy at T_c)	$-2^{\frac{1}{2}}$	-2	$-2/3^{\frac{1}{2}}$
S_c/Nk (Entropy at T_c)[b]	0.30647	0.33028	0.26471

[a]L. Onsager has noted that for these 3 lattices, arctan $(\sinh 2K_c) = \pi/z$, with z = coordination number (4,6,3)
[b]Note: $S_\infty/Nk = \ln 2 = 0.69315$

3.10 Spontaneous Magnetization and Magnetic Susceptibility

There are compelling reasons for identifying the magnetization with long-range order (LRO). In molecular-field theory, they are indistinguisable. In *any* theory based on spin variables,

$$M_z^2(T) = \frac{1}{N} \left\langle \left(\sum_i S_i^z \right)^2 \right\rangle_{TA} = \lim_{R_{ji} \to \infty} <S_i^z S_j^z>_{TA} \quad . \tag{3.10.1}$$

Divison by the total number of spins N in the thermodynamic limit eliminates all but the asymptotic correlations.

The susceptibility, on the other hand, is related to the manner in which the correlations decay to the asymptotic value. Specifically,

$$\chi_{zz}(T) = \frac{\beta}{N} \left\langle \left[\sum_i S_i^z - M_z(T) \right]^2 \right\rangle_{TA} = \beta \sum_j <S_i^z [S_j^z - M_z(T)]>_{TA} \tag{3.10.2}$$

indicating by the subscript zz that this expression stands for $-\partial^2 f/\partial B_z^2 |_{B=0}$. One derives similar expressions for mixed derivatives, e.g., $-\partial^2 f/\partial B_z \partial B_x |_{B=0}$ and labels them analogously. The transverse susceptibility discussed in Sect.3.6 is χ_{xx}.

By calculating correlation functions such as the above, we can dispense with the magnetic field in the transfer matrix, and make use of the results obtained so far. Actually, in *numerical* calculations of the largest eigen-vaue of the transfer matrix, or in certain series calculations, an external field speeds up the convergence. Unfortunately, it introduces great complexity

into the Jordan-Wigner transformation. We choose distant spins in (3.10.1)
along the same column, avoiding sandwiching transfer matrices between the
spins, cf. (3.5.25,27. We must follow the transformations through which we
diagonalized $\mathbf{V}$, starting with the substitution of S_i^z by the Pauli matrix
$\sigma_i^z \rightarrow \sigma_i^x$, thence to fermions and finally to normal modes (eigenstates of $\mathbf{V}$).
It is convenient to use an identity, (3.6.8), [Ref.3.11, Eq. (3.100)]:

$$\exp(i\pi n_m) = (c_m^* + c_m)(c_m^* - c_m) \tag{3.10.3}$$

to obtain a compact expression for the correlations among the S_i^z of (3.8.7):

$$S_i^z S_j^z \Rightarrow (c_i^* - c_i)(c_{i+1}^* + c_{i+1})(c_{i+1}^* - c_{i+1})\cdots(c_{j-1}^* - c_{j-1})(c_j^* + c_j)$$

and

$$\langle S_i^z S_j^z \rangle_{TA} \Rightarrow (0|(c_i^* - c_i)\cdots(c_j^* + c_j)|0) \quad . \tag{3.10.4}$$

The thermal average is just the ground-state expectation value of this expression, taking care that there may be two ground states! It is easily calculable by Wick's theorem, which states: "associate the operators in pairs
in their original order, replace each pair by its ground state expectation,
multiply the product of these pairs by $(-)^P$ (where$(-)^P$ is the parity of the
permutation required to bring the paired operators back into the original
ordering) then sum over all possible distinct such pairings". We distinguish
three possible pairings, but the only *nonvanishing-type pairing* takes the
form:

$$a_{ij} = (0|(c_i^* - c_i)(c_{j+1}^* + c_{j+1})|0) \quad . \tag{3.10.5}$$

With the inverse of the Bogoliubov transformation being

$$a_q = b_q \cos\theta_q - b_{-q}^* \sin\theta_q \tag{3.10.6}$$

(with θ_q given in (3.9.14) odd in q), and (3.9.4) giving the c_i's in terms
of the a_q's, we express (3.10.5) in terms of the b_q and b_q^* which are the
exact lowering and raising operators of the transfer matrix. We find

$$c_n^* \pm c_n = \frac{1}{N^{\frac{1}{2}}} \sum_q e^{iqn}[b_q(\sin\theta_q \pm \cos\theta_q) + b_{-q}^*(\cos\theta_q \mp \sin\theta_q)] \tag{3.10.7}$$

and by virtue of

$$(0|b_q b_{q'}^*|0) = \delta_{q,q'} \quad , \tag{3.10.8}$$

we evaluate the factors a_{ij},

$$a_{ij} = a(j - i) = -\frac{1}{N} \sum_q \exp[-iq(j - i)] \exp[-i(2\theta_q + q)] \quad . \tag{3.10.9}$$

Problem 3.14. (a) Fill in the details leading to (3.10.9) above.

(b) Prove that the contractions such as the following, vanish:

$$(0|(c_n^* \pm c_n)(c_m^* \pm c_m)|0) = 0 \quad \text{for } n \neq m \tag{3.10.10}$$

by using (3.10.7,8).

(c) Obtain their value for $n = m$ by the same methods, then confirm the results by transforming back to Pauli spin operators. (*Hint*: $\sigma_x^2 = 1$)

. . . .

Using Wick's theorem to obtain the ground-state expectation of (3.10.4), we take a product of the a_{ji}'s plus all the signed permutations $\mathbf{P}$:

$$<S_i^z S_j^z>_{TA} \Rightarrow a_{ii} \, a_{i+1,i+1} \cdots a_{j-1,j-1} + \sum_p^P (-)^P \mathbf{P}(a_{ii} a_{i+1,i+1} \cdots a_{j-1,j-1}) \; .$$

Only the contractions of type (3.10.5) contribute, the other possible contractions vanishing according to (3.10.10). The result is, by inspection, a determinant:

$$<S_i^z S_j^z>_{TA} = \begin{Vmatrix} a_{ii} & a_{i,i+1} & \cdots & a_{i,j-1} \\ & a_{i+1,i+1} & \cdots & \\ & & & \\ \cdot & & \cdot & \cdot \\ \cdot & & \cdot & \cdot \\ \cdot & & \cdot & \cdot \\ a_{j-2,i} & \cdots & \cdot & \cdot \\ a_{j-1,i} & & & a_{j-1,j-1} \end{Vmatrix} \tag{3.10.11}$$

the elements a_{ik} of which depend only on the distance from the main diagonal, $i-k$. Such determinants, known as *Toeplitz* determinants, differ only slightly from the corresponding *cyclic* determinants, although this difference can be significant as we see in the example below.

Example:

$$\text{Toeplitz:} \quad \begin{Vmatrix} 010000\ldots \\ 0010000.. \\ 00010000. \\ \cdot\cdot\cdot\cdot\cdot \\ \cdot\cdot\cdot\cdot\cdot\cdot\cdot \\ \cdot\cdot\cdot\cdot\cdot\cdot \\ 000000000 \end{Vmatrix} = 0 \; ,$$

whereas

$$\text{Cyclic:} \quad \begin{Vmatrix} 010000\ldots \\ 0010000.. \\ \cdot\cdot\cdot\cdot\cdot\cdot\cdot\cdot\cdot \\ \cdot\cdot\cdot\cdot\cdot\cdot\cdot\cdot \\ 100000000 \end{Vmatrix} = 1$$

for all dimensionality.

Happily, there exists a relevant theorem by Szegö and Kac (which might even have originated with *Onsager* [3.21]) relating the Toeplitz form to the easily calculated cyclic determinant, with an accuracy $O(1/|j-i|)$. (This accuracy is just sufficient for the calculation of the magnetization but insufficient for the calculation of the susceptibility, which is perhaps one reason for the ongoing research into Toeplitz forms.) Briefly, the Kac-Szegö theorem states that the Toeplitz determinant T is proportional to the analogous cyclic determinant C as follows

$$T = C \, \exp \left(\sum_{n=1}^{\infty} n k_n k_{-n} \right) \quad . \tag{3.10.12}$$

The k's are obtained from the eigenvalues $F(q)$ of the cyclic matrix,

$$F(q) \equiv \sum_m e^{iqm} a(m) \tag{3.10.13}$$

by the relation

$$\ln F(q) = \sum_{n=-\infty}^{+\infty} k_n \, e^{inq} \quad . \tag{3.10.14}$$

Finally, the cyclic determinant is just the product,

$$C = \prod_q F(q) = \exp \left[|j-i| \int_{-\pi}^{+\pi} \frac{dq}{2\pi} \ln F(q) \right] \tag{3.10.15}$$

The second expression being valid in the limit $|j-i| \to \infty$.

With the a(m) given in (3.10.9), we find F by inspection:

$$F(q) = \exp[i(2\theta_q + q)] \quad \text{and} \quad \ln F(q) = i(2\theta_q + q) \quad . \tag{3.10.16}$$

As q and θ_q are both odd functions, *the exponent in C* (3.10.15) *vanishes* and $C = 1$. To obtain the k_n's, we must do more algebra, solving first for $\exp(-2i\theta_q)$ in (3.9.14,15) which is found to be:

$$\exp(-2i\theta_q) = e^{iq} \left[\frac{(1 - x_1^{-1} e^{iq})(1 - x_2 \, e^{-iq})}{(1 - x_1^{-1} e^{-iq})(1 - x_2 \, e^{iq})} \right]^{\frac{1}{2}} \quad (T < T_c) \tag{3.10.17a}$$

and

$$\exp(-2i\theta_q) = - \left[\frac{(1 - x_1^{-1} e^{iq})(1 - x_2^{-1} e^{iq})}{(1 - x_1^{-1} e^{-iq})(1 - x_2^{-1} e^{-iq})} \right]^{\frac{1}{2}} \quad (T > T_c) \tag{3.10.17b}$$

in which

$$\begin{aligned} x_1 &= \coth K_1^* \coth K_2 > 1 \quad \text{and} \\ x_2 &= \coth K_2 \tanh K_1^* \gtrless 1 \text{ for } T \gtrless T_c \quad . \end{aligned} \tag{3.10.18}$$

140

The phase angles in all four factors in (3.10.17a or b) are taken in the interval $-\frac{1}{2}\pi$, $+\frac{1}{2}\pi$.

Next, we solve (3.10.14) for k_n, using (3.10.17a):

$$k_n = \frac{i}{2\pi} \int_{-\pi}^{+\pi} dq\, e^{-inq} (2\theta_q + q)$$

$$= \frac{1}{4\pi} \int_{-\pi}^{+\pi} dq\, e^{-inq} \left[\ln(1 - x_1^{-1} e^{-iq}) - \ln(1 - x_1^{-1} e^{iq}) \right.$$

$$\left. + \ln(1 - x_2 e^{+iq}) - \ln(1 - x_2 e^{-iq}) \right] . \tag{3.10.19}$$

Because of the inequalities (3.10.18) the logarithms are expandable in a Taylor series:

$$\ln(1 - x) = - \sum_{n=1}^{+\infty} \frac{x^n}{n} ,$$

so that the above integrations are easily performed, and we obtain

$$k_n = \frac{1}{2} \frac{1}{n} \left(x_1^{-|n|} - x_2^{|n|} \right) , \qquad \text{hence} \tag{3.10.20}$$

$$\sum_{n=1}^{\infty} n k_n k_{-n} = -\frac{1}{4} \sum_{n=1}^{\infty} \frac{1}{n} \left[x_1^{-2n} + x_2^{2n} - 2(x_2 x_1^{-1})^n \right]$$

$$= \frac{1}{4} \ln \left[(1 - x_1^{-2})(1 - x_2^2)(1 - x_2 x_1^{-1})^{-2} \right] . \tag{3.10.21}$$

When this expression is inserted into the exponential in (3.10.12) it yields the correlation function,

$$\lim_{|j-i| \to \infty} \langle S_i^z S_j^z \rangle_{TA} = T = \left(1 - \frac{1}{\sinh^2 2K_1 \sinh^2 2K_2} \right)^{\frac{1}{4}} \tag{3.10.22}$$

for $T \leqslant T_c$ [note: T_c is given by $\sinh(2K_{1c}) \sinh(2K_{2c}) = 1$]. As an obvious consequence, the spontaneous magnetization $m(T)$ at $T \leqslant T_c$ is

$$m(T) = |M_z(T)| = \left(1 - \frac{1}{\sinh^2 2K_1 \sinh^2 2K_2} \right)^{1/8} \tag{3.10.23}$$

which can be interpreted crudely as $m \propto (T_c - T)^{1/8}$ (in sharp contrast with MFT 1/2 power dependence).

None of these calculations depended on whether the vacuum had even or odd occupation, therefore the conclusions are general.

. . . .

Problem 3.15. Using the appropriate form (3.10.17b), show $m = 0$ *above* T_c, in the manner of (3.10.19-23).

. . . .

The isotropy of the above results is somewhat amazing. Because we obtain for the correlations along a column a function which is symmetric in J_1 and J_2, we will find the same along a row (by rotating the direction of propagation of the transfer matrix $90°$). This implies that long-range order is truly isotropic-independent both of distance and of angle relative to the crystal axes in the large distance limit. The magnetization is thus isotropic and uniform.

The magnetic susceptibility is a more complicated matter. As early as 1951, *Oguchi* [3.22] showed that the zero-field susceptibility was a power series in the $w_{ij} = \tanh\beta J_{ij}$, and obtained for the nearest-neighbor model ($J_{ij} = J$ if i and j are nearest neighbors, $=0$ otherwise),

$$\chi = \frac{\mathfrak{C}_{\frac{1}{2}}}{T}\left(\sum_r a_r w^r\right) \equiv \frac{\mathfrak{C}_{\frac{1}{2}}}{T}[X] \qquad (3.10.24)$$

where $a_0 = 1$ and $a_r = $ twice the term linear in N in the total number of ways of placing a graph of r lines on the lattice of N sites, such that all but two of the vertices are the meeting points of an even number of lines. This first theory of magnetic graphs can be derived on the basis of (3.1.8), but we shall not dwell on it. There have been a number of attempts to calculate X, mainly by numerical techniques.

M.F. Sykes has used bond counting techniques to obtain a large number of terms in this expansion, for various lattices in 2D and 3D. His calculations reveal the first 15 terms for the sq lattice [3.23]:

$$X = 1 + 4w + 12w^2 + 36w^3 + 100w^4 + 276w^5 + 740w^6 + 1972w^7$$

$$+ 5172w^8 + 13492w^9 + 34876w^{10} + 89764w^{11}$$

$$+ 229628w^{12} + 585508w^{13} + 1486308w^{14} + 3763460w^{15}\ldots . \qquad (3.10.25)$$

Unlike our primitive series in (3.1.5) which contained just the first two terms in this expansion, there is sufficient information in this series to unambiguously fix an exponent γ in the expression,

$$X \sim \left(1 - \frac{T_c}{T}\right)^{-\gamma} \qquad (3.10.26)$$

by curve-fitting and numerical analysis. The result is

$$\gamma = 7/4 \quad (\pm 10^{-2} \text{ or better}) \quad . \qquad (3.10.27)$$

This critical exponent is found to fit the Ising ferromagnet on *all* principal lattices in two dimensions. In 3D, a similar calculation yields $\gamma = 5/4$ on all type lattices. (We expect that as the number d of dimensions is increased further, the MFT result $\gamma = 1$ is approached at $d \to \infty$. In fact, there is good reason to believe that the MFT limit is achieved already at $d = 4$ or 5, as for the spherical model!) These inferences were confirmed subsequently by the powerful Padé approximant techniques of *Baker* [3.24] and, *in 2D*, by a theoretical analysis of the asymptotic behavior of the correlation function. (We describe this work by *Barouch* et al. [3.25] in the briefest manner. After establishing that the correlation function in (3.10.2) decays as

$$R^{-\frac{1}{4}} F_1(t) + R^{-5/4} F_2(t) + O(R^{-9/4+0^+})$$

with R a temperature-dependent length related to the actual distance $|j - i|$, and replacing sums by integrals (in the critical region where this is permissible) they obtain the coefficients in the expression, valid *near* T_c:

$$X = C_{0\pm} |1 - T_c/T|^{-7/4} + C_{1\pm} |1 - T_c/T|^{-3/4} + O(1)$$

as $C_{0-} = 0.0255...$, $C_{0+} = 0.9625...$, $C_{1-} = -0.00199...$ and $C_{1+} = 0.07499...$, where $\pm$ is for $T \gtrless T_c$. With the corrections $C_{1\pm}$ small, it certainly seems that the critical phenomena are well represented by the leading singularity in this instance.)

At the point T_c where χ diverges but the spontaneous magnetization still vanishes, it is reasonable to expect that $m(T_c, B)$ depends sublinearly on B, e.g.,

$$|m| = A|B|^{1/\delta} , \quad \text{with} \quad \delta > 1 . \tag{3.10.28}$$

The exponent δ is an important critical exponent. Although it is impossible to diagonalize the infinite transfer matrix in the presence of an external field by the methods developed in this chapter, the numerical evaluation of the largest eigenvalue for a finite two-dimensional strip converges quickly and yields accurate prediction. In particular, $\delta = 15$ for the 2D Ising model is obtained from a fit of the data to the formula [3.26]:

$$|m|^{15} = |\tanh B/kT_c| \tag{3.10.29}$$

which describes the approach to saturation qualitatively well, and the small-field behavior precisely.

3.11 Zeros of the Partition Function

Lee and *Yang* discovered a remarkable property of the Ising model, and of the analogous lattice gas as well [3.6]. Briefly, by allowing the quantity -2B/kT to be complex, and determining the values of this quantity at which Z vanishes they were able to infer the essential analytic properties of the free energy. Defining the complex variable ζ,

$$\zeta = e^{-2B/kT} \tag{3.11.1}$$

they showed the partition function for the N spins of an Ising ferromagnet to be a polynomial in ζ of degree N, the zeros of which all lie on the unit circle $|\zeta| = 1$. In the thermodynamic limit the distribution of zeros is therefore a continuous T-dependent function $g(\theta)$, where θ is the angle of ζ in the complex plane. From this observation follow several interesting results. First, the spontaneous magnetization, $m(T)$, is

$$m = 2\pi g(0) \quad . \tag{3.11.2}$$

Generally, the magnetization in finite (complex) field is given by an integral,

$$M(\zeta) = 1 - 4\zeta \int_0^\pi g(\theta) \, \frac{\zeta - \cos\theta}{\zeta^2 - 2\zeta\cos\theta + 1} \, d\theta$$

$$= 1 - 2\zeta \int_{-\pi}^\pi g(\theta)(\zeta - e^{i\theta})^{-1} \, d\theta \quad . \tag{3.11.3}$$

The free energy per spin is, with E_0 the ground state energy,

$$f = \frac{E_0}{N} - B + kT \int_0^\pi g(\theta) \ln (\zeta^2 - 2\zeta\cos\theta + 1) d\theta \quad . \tag{3.11.4}$$

The function $g(\theta) = -g(-\theta)$ is symmetric, positive, and is normalized such that the integral around the unit circle is 1, or

$$\int_0^\pi g(\theta) \, d\theta = \frac{1}{2} \quad . \tag{3.11.5}$$

Thus, it is the temperature-dependence of $g(\theta)$ which yields the interesting temperature dependence of f and M, and $g(T,\theta)$ can itself be considered a thermodynamic function of considerable importance. With it, one can compute the properties of an Ising ferromagnet in finite field, in the thermodynamic limit ($N \rightarrow \infty$) where all previous non-extrapolatory methods had failed. Unfortunately, to this date $g(\theta)$ is known only in 1D. (Another clear challenge to our undaunted reader!)

We now derive the above formulas and their applications as simply as possible. We consider an Ising ferromagnet in which all nonzero bonds (not necessarily nearest-neighbor) are ferromagnetic in sign. Denote the corresponding Hamiltonian H_0, the interaction with an external field being $-B\sum_i S_i$. As usual, the $S_i = \pm 1$. With the trivial identity

$$\exp[B(S_i - 1)/kT] = \frac{1}{2} (1 + \zeta) + \frac{1}{2} (1 - \zeta)S_i \tag{3.11.6}$$

we write the partition function in the form

$$Z = \exp(NB/kT) \exp(-E_0/kT) \, Q \tag{3.11.7}$$

where E_0 is the energy of H_0 when all spins are up (all $S_i = +1$) and

$$\begin{aligned}
Q &= \mathrm{Tr} \left\{ \exp[-(H_0 - E_0)/kT] \prod_i \exp[B(S_i - 1)/kT] \right\} \\
&= \sum_{n=0}^{N} P_n \zeta^n \quad .
\end{aligned} \tag{3.11.8}$$

It is easily verified that the coefficients $P_n = P_{N-n}$ are real, positive, and $P_0 = P_N = 1$. Clearly, knowledge of the above polynomial in ζ is tantamount to calculation of the partition function.

The Lee-Yang theorem that all the zeros of Z (hence of Q) lie on the unit circle of complex ζ is relatively lengthy to prove, although it is quite plausible. In the absence of H_0, Q is easily seen to be just $(1+\zeta)^N$, and the P_n are the ordinary binomial coefficients. The zeros all lie at $\zeta = -1$. Introducing an interaction between 2 spins S_1 and S_2 and no others, Q becomes $(1+\zeta)^{N-2}(1 + 2\zeta e^{-2K} + \zeta^2)$, where $K = J/kT$ as usual. Now, $N-2$ zeros lie at -1 and 2 lie at conjugate points:

$$\zeta_{\pm} = -e^{-2K} \pm i(1 - e^{-4K}) \quad .$$

. . . .

Problem 3.16. With 4 spins interacting: $H_0 - E_0 = -\frac{1}{2} J(S_1 + S_2 + S_3 + S_4)^2 + 8J$ locate the zeros of the partition function on the unit circle and their angles, showing they are symmetrically distributed about $\theta = 0$.

. . . .

There is no spontaneous magnetization and no phase transition unless the zeros pinch the real axis (near $\theta = 0$). Thus it is interesting and important to know the distribution of zeros. The Lee-Yang theorem [3.6] requires a number of lemmas and a proof by induction, which are not reproduced here. An important observation, which limits the applications to ferromagnets, is that the more interactions we introduce, the smaller the coefficients P_n become compared to the binomial coefficients; and the closer n is to $\frac{1}{2} N$, the greater is this depression. In the limit of very low temperatures, where the factors $\exp(-2K)$ are essentially zero,

$$\lim_{T \to 0} Q = 1 + \zeta^N + O(e^{-2K}) \quad ,$$

the points $\exp(i\theta_r)$ with $\theta_r = (2r+1)\pi/N$, r an integer in the interval $-\frac{1}{2}N$, $+\frac{1}{2}N$, mark the vanishing of Q. In this example, g would be a constant $1/2\pi$, and the spontaneous magnetization $m = 1$ according to (3.11.2). In the op-

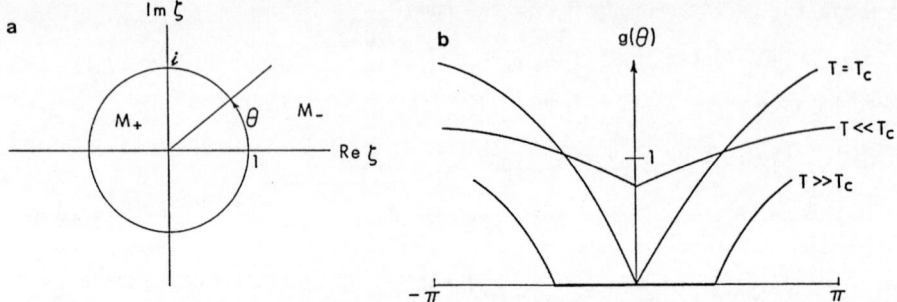

Fig.3.10. (a) Complex $\zeta = \exp(-2B/kT)$ plane. M_- is analytic outside unit circle, M_+ within it. The discontinuity $M_+ - M_-$ is related to $g(\theta)$ by (3.11.9). The behavior of $g(\theta)$ with θ is shown in (b) for various temperature

posite limit of no interactions, we have seen all the zeros are at -1. Raising the temperature *generally* moves the distribution from the first and fourth quadrants into the second and third quandrants of the unit circle, as indicated in the Fig.3.10. Thus, spontaneous magnetization always disappears at high T.

Once we know the N zeros are on the unit circle, we can denote them $\exp(i\theta_k)$, and $Q = \prod_k [\zeta - \exp(i\theta_k)]$ by definition. Taking logarithms and using the symmetry of $g(\theta)$ establishes (3.11.4). The formula for M then follows by differentiation with respect to B. While these are straightforward, the simple result (3.11.2) for the spontaneous magnetization requires an explanation. For ζ on the positive real axis outside the unit circle M is real but negative. Inside the unit circle, it is real and positive, for ζ on the real axis in the interval 0,1. The discontinuity at $\zeta = 1$ is obtained from the branch cut in (3.11.3).

The function in (3.11.3) has a discontinuity everywhere on the unit circle, not merely on the real axis. If we define M_- to be this function outside the unit circle and M_+ the same function inside of it, and also define the discontinuity as $2M_0(\phi) = M_+(\phi) - M_-(\phi)$ at $r = 1$, we then calculate it to have the value,

$$M_0(\phi) = 2\pi \, e^{i\phi} \, g(\phi) \quad . \tag{3.11.9}$$

This relation generalizes (3.11.2) and identifies $g(\phi)$ with the thermodynamic function $M_0(\phi)$. But it should be noted that M_0 is the usual magnetization m(T) only at $\phi = 0$, elsewhere it can even exceed 1. In general, there has been no identification of $g(\phi)$ or $M_0(\phi)$ with a suitable correlation function, although this would mark useful progress.

We proceed with an example, and find the zeros of the one-dimensional Ising chain analyzed in Sect.3.5. From (3.5.13),

$$\lambda_\pm = e^{(B+J)/kT}\left\{\frac{1}{2}(1+\zeta) \pm \left[\frac{1}{4}(1-\zeta)^2 + \zeta e^{-4K}\right]^{\frac{1}{2}}\right\} \tag{3.11.10}$$

are the eigenvalues of the transfer matrix, in the new language using ζ.

Z will vanish if

$$\lambda_+^N + \lambda_-^N = 0 \quad , \tag{3.11.11}$$

therefore we solve for $\lambda_- = \lambda_+ \exp(i\pi p/N)$, with p any odd integer in the interval $-N, +N$. After elementary algebra, we obtain the roots at $z_p \equiv \exp(i\theta_p)$:

$$z_p = - e^{-4K} + (1 - e^{-4K}) \cos \frac{\pi p}{N} \pm i \left\{1 - \left[- e^{-4K} + (1 - e^{-4K}) \cos \frac{p}{N}\right]^2\right\}^{\frac{1}{2}} \tag{3.11.12}$$

This shows $g(\theta)$ is symmetric, and allows us to compute it in the limit $N \to \infty$:

$$g(\theta) = \frac{1}{2\pi} \frac{\sin \frac{1}{2}\theta}{(\sin^2 \frac{1}{2}\theta - e^{-4K})^{\frac{1}{2}}} \tag{3.11.13}$$

for θ in the interval $[-\pi, \cos^{-1}(1 - 2 e^{-4K})]$ and $g = 0$ for θ in the interval $[0, \cos^{-1}(1 - 2 e^{-4K})]$.

··

Problem 3.17. Show that with this formula for $g(\theta)$ the integral (3.11.4) will yield the correct free energy $-kT \ln\lambda_+$ in 1D. Derive

$$M(\zeta) = \left(\frac{\zeta^2 - 2\zeta + 1}{\zeta^2 - 2\zeta(1 - 2 e^{-4K}) + 1}\right)^{\frac{1}{2}} \quad , \tag{3.11.14}$$

and use this to compute the *susceptibility*, verifying (3.5.17).

··

Terminating this section, it is appropriate to mention that the solution of the 2D Ising model in a magnetic field has been reformulated by *Barouch* [3.27], who produced an exact formula for Z in terms of the —alas unknown! — distribution of prime numbers. We list 2 other applications [3.27].

3.12 Miscellania, Including 2D Antiferromagnets

The dual of the triangular lattice is the honeycomb (hexagonal), and vice versa, as we saw in Fig.3.3. This duality cannot immediately locate the critical point of either structure, nor map the high temperature properties onto the low temperature properties for a given lattice in a direct way. On-

147

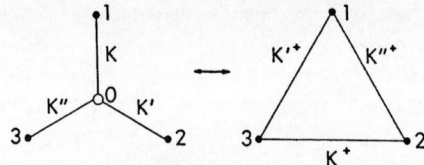

Fig.3.11. Star-Triangle transformation. Its application to stars on the H lattice (o oo and their 3 neighbors) leads to the T lattice

sager invented the *star-triangle* transformation to relate the partition functions on these two lattices, thus determining them both. This methodology has recently been used by *Baxter* and *Enting* in a paper amusingly entitled "399th solution of the Ising model" to connect the two abovementioned lattices to the square, and then to obtain the partition functions of all of them without use of transfer matrices, series summations, or other special tricks [3.28]. The star-triangle transformation is essential in the study of the triangular lattice, on which one has an antiferromagnet with very unusual properties. We now derive its properties, following Domb's exposition in [Ref.3.12, p. 182ff].

Figure 3.11 illustrates the basic star and triangle in question. The star contains an extra spin, so it is the partial trace of the former that will be equal to the Boltzmann factor for the latter. We sum over S_0 in the star:

$$\sum_{S_0=\pm 1} \exp[KS_0(S_1 + S_2 + S_3)] = 2\cosh^3 K + 2(\sinh^2 K)$$

$$\times (\cosh K)(S_1 S_2 + S_2 S_3 + S_3 S_1) \tag{3.12.1}$$

and denoting the parameter for the triangle as K^+, we can compare the above with

$$\exp[K^+(S_1 S_2 + S_2 S_3 + S_3 S_1)] = (\cosh^3 K^+ + \sinh^3 K^+)$$

$$+ \frac{1}{2} e^{K^+} (\sinh 2K^+)(S_1 S_2 + S_2 S_3 + S_3 S_1) \quad . \tag{3.12.2}$$

The two expressions will agree if the second is multiplied by

$$f = 2\cosh^3 K / (\cosh^3 K^+ + \sinh^3 K^+) \tag{3.12.3}$$

and if we choose K^+ to satisfy

$$\frac{\frac{1}{2}e^{K^+}\sinh(2K^+)}{\cosh^3 K^+ + \sinh^3 K^+} = \tanh^2 K \quad . \tag{3.12.4}$$

Now applying the star-triangle transformation to the open circles in Fig. 3.12 we readily transform the H into the T lattice, with H: honeyc., T: triang.;

$$Z_{2N}^{(H)}(K) = f^N Z_N^{(T)}(K^+) \quad . \tag{3.12.5}$$

The duality relationships (3.3.2-7) relate properties on one lattice at K with those on the other at K^*:

$$\frac{Z_N^{(T)}(K^*)}{2^{\frac{1}{2}N}(\cosh 2K^*)^{3N/2}} = \frac{Z_{2N}^{(H)}(K)}{2^N(\cosh 2K)^{3N/2}} \quad . \tag{3.12.6}$$

Simplifying (3.12.4) we find

$$e^{4K^+} = 2 \cosh 2K - 1 \tag{3.12.7}$$

and with the help of (3.3.8),

$$(e^{4K^+} - 1)(e^{4K^*} - 1) = 4 \quad . \tag{3.12.8}$$

The critical point on the triangular (T) lattice is at $K^+ = K^*$, i.e., at $\exp(4K_c^+) = 3$, as recorded in Table 3.2. That of the H lattice is at $K = (K^+)^*$, which, after some algebra, is located at $\exp(2K_c) = 2 + \sqrt{3}$ also given in the table. For details, see the original work of *Wannier* [3.29] or the review [3.12] or below in the present section, where formulas for anisotropic lattices (different J's in different directions) are derived and used.

The critical point for the sq ferromagnetic Ising model was given as $\sinh^2 2K_c = 1$ in Sect.3.3. It is evident that for antiferromagnetic coupling on the same lattice, i.e., $J < 0$, the critical temperature now called T_N is given by *precisely the same expression*. In the absence of an external field, a symmetry operation (reversing every second spin) effectively reverses the sign of J without affecting the partition function.

This is patently untrue for the T lattice. For when J is negative, K^+ is negative and there is no K with which one can satisfy (3.12.7)! Indeed, there is no phase transition for the isotropic T antiferromagnet, which is in many other respects a remarkable system. Its ground state has energy 1/3 that of the corresponding ferromagnet and is so degenerate —that is, there are so many configurations having this energy —that the high-temperature disordered phase persists down to $T = 0$. Now, it is obvious that within each elementary triangle, the optimum energy is obtained for 2 spins of one sign and the 3rd spin of the opposite sign. The energy of such a state —one-third the ground state energy of the corresponding ferromagnet— is thus a *lower* bound. But, starting with a single cell in this configuration and working outward, one may construct such configurations in every triangle of the plane. We have this as a variational ground state, i.e., as an upper bound and therefore it must be the exact ground state energy. As this ground state is enormously degenerate, $O(2^{pN})$ with $0 < p < 1$, it is desired to obtain the ground state entropy. *Wannier*, first to investigate this remarkable antiferromagnet [3.29] obtained the exact transfer matrix and was able to establish the formula for the ground-state entropy:

$$\mathscr{S}_0/Nk = \frac{3}{\pi} \int_0^{\pi/6} d\theta \ \ln(2\cos\theta) = 0.32306\ldots \ . \tag{3.12.9}$$

This number is approximately equal to the entropy $\mathscr{S}_c$ of the corresponding ferromagnet at its (nonzero) T_c. In Table 3.2 we observe that the 2D Ising model usually has a *critical* entropy of this magnitude, approximately half the maximum value (ln2). In a few pages, we shall see that Wannier's antiferromagnet has, indeed, its phase transition at $T = 0$.

A large number of lattices have been studied in 2D, as listed in [3.12]. Their study always follows the same routine which we illustrate for the sq, T and H lattices (following *Wannier* [3.29] as generalized by *Utiyama* [3.30]). A transfer matrix is derived, the solution of which (by a transformation to free fermions) yields all the thermodynamic properties.

The difficulty with the straightforward approach in the case of arbitrary lattices, is that the transfer matrix is not now one-to-one. A single spin on the nth row connects to 2 spins on the $n+1^{st}$ row, and one cannot express this by a simple $\mathbf{V}^x$ matrix as in (3.8.1). But if we introduce extra spins to restore rectangular symmetry (or almost) and then freeze them in or out by setting certain of the coupling constants equal to infinity (or zero), we may achieve the desired results. There is no unique way to do this; however Fig.3.12 contains the three principal lattices as special cases for the indicated values of the J's. The spins have to be moved somewhat (without disturbing the connectivity or crossing any of the lines) to display the desired T or H geometries explicitly.

If we wish to establish the general transfer operator for the grid illustrated in Fig.3.12, then specialize to the indicated J's to calculate the properties of the three special lattices, then we should transfer from the nth column to the $n+2^{nd}$, or from the $n+1^{st}$ to the $n+3^{rd}$, alternate columns (or rows) being distinct in general. As we must include two columns in the

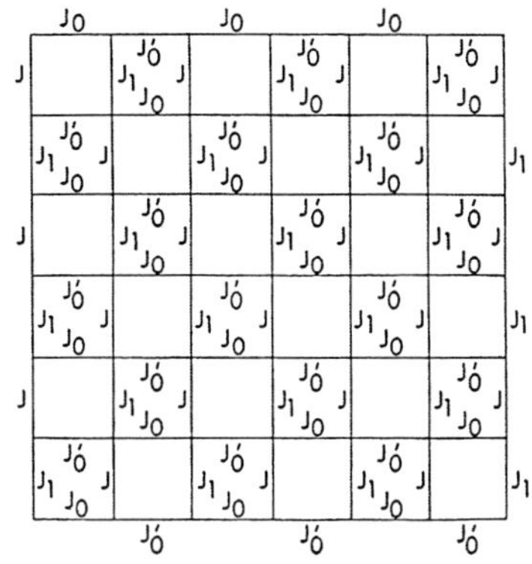

Fig.3.12. The checker lattice beomces the sq ($J = J_1$, $J_0 = J_0'$). the hexagonal ($J = 0$) and triangular ($J = \infty$)

operator, the eigenvalue will be λ^2. We then solve

$$\mathbf{W} \cdot \mathbf{u} = \lambda^2 \mathbf{u} \tag{3.12.10}$$

where the following $\mathbf{W}$ replaces $\mathbf{V}^x$, $\mathbf{V}^y$ and $\mathbf{V}^B$ of (3.9.1):

$$\mathbf{W} = A\mathbf{V}_1\mathbf{V}_2\mathbf{V}_3\mathbf{V}_4\mathbf{V}_5\mathbf{V}_6 \quad . \tag{3.12.11}$$

With

$$A = (2 \sinh 2K_0)^{\frac{1}{2}N} (2 \sinh 2K_0')^{\frac{1}{2}N} \text{ is a constant,}$$

$$\mathbf{V}_1 = \exp\left[-\left(K_0^* \sum \sigma_{2j}^z + K_0'^* \sum \sigma_{2j+1}^z \right) \right]$$

$$\mathbf{V}_2 = \exp\left(K \sum \sigma_{2j}^x \sigma_{2j+1}^x + K_1 \sum \sigma_{2j+1}^x \sigma_{2j+2}^x \right)$$

$$\mathbf{V}_3 = \exp\left(h \sum \sigma_j^x \right)$$

$$\mathbf{V}_4 = \exp\left[-\left(K_0^* \sum \sigma_{2j+1}^z + K_0'^* \sum \sigma_{2j}^z \right) \right]$$

$$\mathbf{V}_5 = \exp\left(K \sum \sigma_{2j+1}^x \sigma_{2j+2}^x + K_1 \sum \sigma_{2j}^x \sigma_{2j+1}^x \right)$$

$$\mathbf{V}_6 = \exp\left(h \sum \sigma_j^x \right) \quad .$$

A solution in closed form will be possible only for $h = 0$, $i\pi/2$ or $i\pi$. For $h = i\pi$, the operators $\mathbf{V}_3$ and $\mathbf{V}_6$ reduce to ± 1 and can be trivially absorbed into the constant. For $h = i\pi/2$, i.e. $\zeta = -1$,

$$\exp\left(\frac{1}{2} i\pi\sigma_j^x \right) = i\sigma_j^x \tag{3.12.12}$$

is a trivial identity which we can use twice:

$$\exp\left(\frac{1}{2} \; i\pi\sigma^x_{2j}\right) \exp\left(\frac{1}{2} \; i\pi\sigma^x_{2j+1}\right) = i \; \exp\left[\frac{1}{2} \; i\pi(\sigma^x_{2j}\sigma^x_{2j+1})\right] \quad . \tag{3.12.13}$$

The product of i's can be absorbed into the constant, and the complex exponentials into V_2 and V_5 by redefining

$$K \rightarrow K + \frac{1}{2} \; i\pi \quad \text{and} \quad K_1 \rightarrow K_1 + \frac{1}{2} \; i\pi \quad . \tag{3.12.14}$$

Subsequently, the calculation proceeds as for real K's and one finds formulas originally given (inscrutably, without explanation,) in [3.6].

To calculate the optimum eigenvalue of (3.12.10) one proceeds almost as for the sq lattice, except that he takes 2 sites (2j, 2j +1) as the unit cell. The transformation to fermions is facilitated if operators on even-numbered sites are denoted c_{2j}, those on odd-numbered sites, b_{2j+1}. For each k, this yields 4×4 matrices resulting from the b and c operators, rather than the 2×2 in the original method of Sect.3.9. Leaving algebraic details aside, we quote the result for the free energy in the *triangular lattice* with totally anisotropic bonds, J,J' and J" in zero field:

$$f = -kT \left\{ \ln 2 + \frac{1}{8\pi^2} \int_{-\pi}^{+\pi} dq_1 \int_{-\pi}^{+\pi} dq_2 \right.$$

$$\times \ln[\cosh 2K \cosh 2K' \cosh 2K'' + \sinh 2K \sinh 2K' \sinh 2K''$$

$$\left. - \sinh 2K \cos q_1 - \sinh 2K' \cos q_2 - \sinh 2K'' \cos(q_1 + q_2)] \right\} \quad .$$
$$\tag{3.12.15}$$

Notice that setting J" = 0 results in the free energy previously computed for the sq lattice, (3.9.21). The phase transition occurs at the value of kT for which [ ] in the above integral just vanishes. For the ferromagnet (all J > 0) this occurs at $q_1 = q_2 = 0$, and it is not difficult to prove that there will always be a solution.

For the *isotropic antiferromagnet* J = J' = J" < 0 discussed in (3.12.9) and *supra*, the breakdown of the star-triangle transformation and the high degeneracy of the ground state seemed to preclude a phase transition. To analyze this point further, let J = J' < 0 and J" ⩽ 0 be independent variable. For |J"| < |J| the minimum of [ ] occurs at $q_1 = q_2 = \pi$. Thus, the conditions for the bracket to vanish at that point determine the critical temperature. The equation is

$$\cosh^2 2K_c \cosh 2K''_c + \sinh 2K''_c(1 - \sinh^2 2K_c) - 2 \sinh 2K_c = 0 \tag{3.12.16}$$

where $K_c = |J|/kT_c$ and $K''_c = |J''|/kT_c$. It has the solution

$$\exp(2K''_c) = \sinh 2K_c \tag{3.12.17}$$

from which we can observe kT_c vanishing as J" approaches J:

$$kT_c \sim 2.89(|J| - |J"|) \quad . \tag{3.12.18}$$

For $|J"| > |J|$ T_c *stays* zero, as there is no possible way for [ ] to vanish, regardless of q_1, q_2. This will be analyzed by the reader in Problem 3.18. Thus, the isotropic AF is a singular case —the *precise* limit at which $T_c \to 0$, and we should not be surprised at the ground state entropy, which equals the typical critical entropy in 2D.

....

Problem 3.18. Study the octant in parameter space $J \leqslant J' \leqslant J" \leqslant 0$ for the triangular AF, showing all the regions where $T_c = 0$.

....

In an *external field* the ground state degeneracies are lifted and there is no reason not to have a phase transition. This situation is interesting, although it resists exact analysis.

The development by *Baxter* and *Enting* [3.28] also relates the three principal lattices in 2D, with the same results as in Fig.3.12 but by a different procedure. These authors use a sequence of star-triangle transformations to functionally transform the honeycomb into sq and T sections, in a manner reminiscent of the work Metamorphose by the great Dutch artist, M. Escher. Their work, compared to his, is reproduced in Fig.3.13, perhaps a new instance of nature imitating art!

We now return to the important topic of the sq *antiferromagnet*. We already know that in the absence of an external field, the thermal properties are identical to those of the corresponding ferromagnet. In an external field, we can take some of the expressions which are accurately known for the ferromagnet and by reversing J (or $w = \tanh J/kT$) obtain the expressions relevant to the antiferromagnet.

Consider the susceptibility as expanded to 15th order in w, (3.10.25). Reversing the sign of J (hence of w) introduces an alternating sign in the expansion which now *vanishes* at a temperature just above T_N. This unphysical behavior due to a failure of convergence has been analyzed by *Fisher* and *Sykes* [3.31], who determined that the susceptibility series should have singularities at both $+w_c$ and $-w_c$. The ferromagnetic singularity $(1 - T_c/T)^{-7/4}$ becomes an innocuous $(1 + T_N/T)^{-7/4}$ in the antiferromagnet. Nevertheless, comparing the expansion of this quantity with the known series leads to tangible results. Supposing X to be of the form:

$$X = (1 + w/w_c)^{-7/4} F(w/w_c) \tag{3.12.19}$$

they determined

$$F = C + D\left(1 - \frac{w}{w_c}\right) \ln\left(1 - \frac{w}{w_c}\right) \qquad (3.12.20)$$

for the AF, with C,D constants to be adjusted for best fit.

It follows that $\chi_{\parallel}$ *peaks* just above T_N, then dropping as the temperature is lowered —with infinite slope at T_N— thence, approaching zero rapidly as $T \to 0$. This behavior differs in detail from that in the AF models studied in Chap.2, namely the MFT in Fig.2.8 and the spherical antiferromagnet in Fig. 2.14. To compare, in Fig.3.14 we show $1/\chi_0$ for the Ising AF and the same quantity for the Ising ferromagnet.

There has been some study of the Ising model AF in finite, homogeneous, external field, with a view to establishing the spin-flop field $B_c(T)$ and the nature of the phase transition across this boundary. The results are not dissimilar from the properties of the simple antiferromagnets considered in Chap.2, except in some details. *Lieb* and *Ruelle* [3.22] have proved a theorem analogous to that of *Lee* and *Yang*, that there is no phase transition in sufficiently weak (real or complex) fields above T_N. Now, at $T = 0$ an external field $B > 2|J|$ does break up the AF ordering in the sq lattice, so an interpolation such as

$$\cosh B_c/kT = \sinh^2 2K \qquad (3.12.21)$$

cannot be too much in error. This is the formula proposed by *Zittartz* and *Müller-Hartmann* [3.33,34] in a somewhat generalized form suitable for anisotropic couplings. Indeed, *Lin* and *Wu* [3.35], probing the validity of such approximations, decided they apply also to the anisotropic T antiferromag-

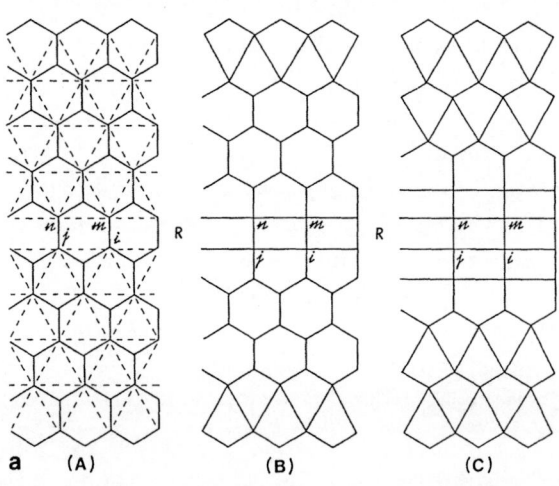

Fig.3.13a,b. Metamorphose: (a) the effects of repeated star-triangle and triangle-star transformations on H lattice [3.28] and (b) the work of M. Escher

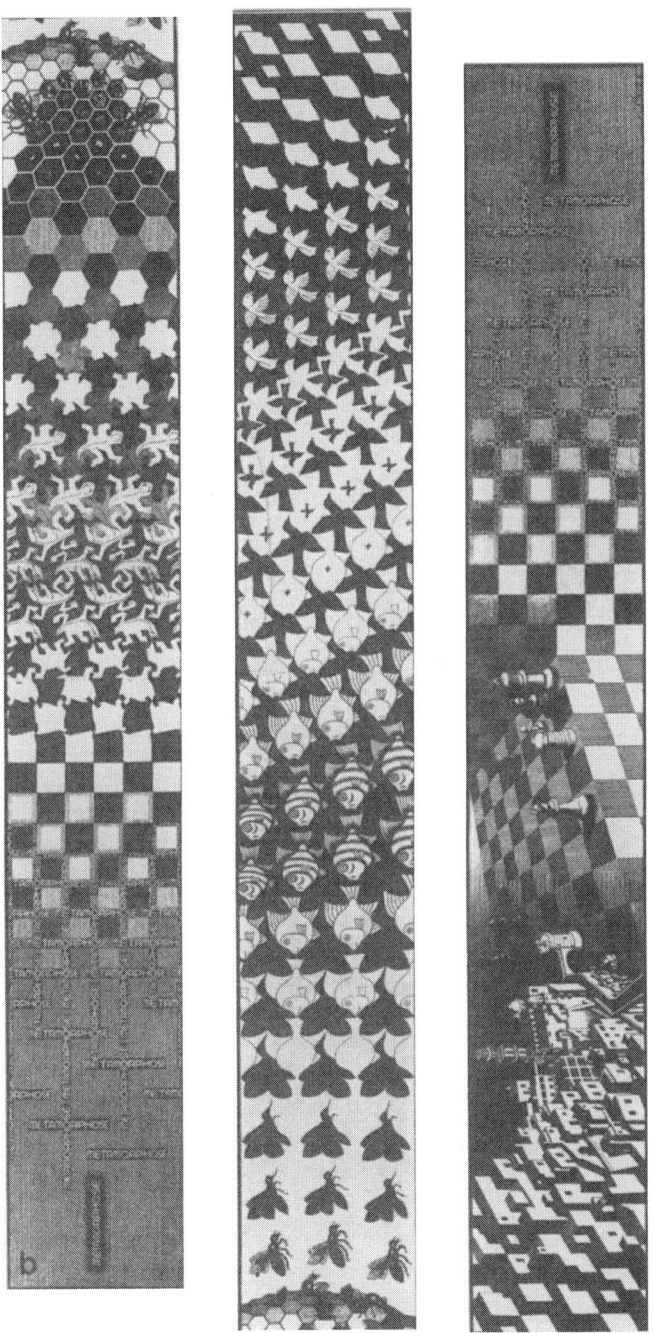

Fig.3.13b (Figure caption see opposite page)

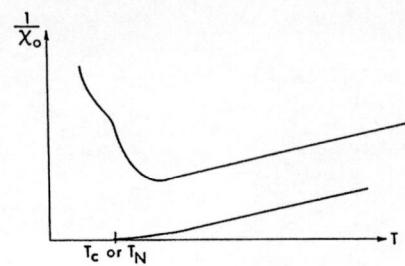

net which, they found, accomodates *several* distinct phases in an applied
field. The original investigation of the AF Ising model in an external
field goes back to *Bienenstock* [3.36], followed by *Plischke* and *Mattis* [3.37]
in connection with Plischke's 1970 doctoral dissertation. His study, which
included next-nearest neighbor interactions, concluded that the logarithmic
specific heat anomaly persists at finite field.

3.13 The Three-Dimensional Ising Model

If this section on the three-dimensional Ising model were limited to exact
results, it would be short indeed. As we shall show below, the agreement
between exact (but numerical) series calculations and experiment is excellent;
but theory —in the form of closed form expressions, relations between physical quantities or manageable approximations —remains sparse, except in the
relatively specialized area of the calculation of critical exponents.

Two antiferromagnets Rb_3CoCl_5 and Cs_3CoCl_5 fullfil the physical require-
ments for the Ising model on a sc lattice (the magnetic atoms, Co, are on a
sc sublattice and their spins are restricted to ±3/2 due to crystal-field
splittings). These substances show specific heat anomalies at their respective values of T_N, as shown in Fig.3.15 [3.38]. Using the series extrapola-
tions and analytical results of *Sykes* et al., *Fisher, Domb,* and their collaborators, *de Jongh* and *Miedema* collected many physically relevant properties
and compared them with experiment [3.39]. We extract from their Table 11
some information on the critical temperatures T_c (or T_N), critical entropy
$\mathscr{S}_c$, high-temperature entropy $\mathscr{S}_\infty$, ground state energy E_0 and critical internal energy U_c. These quantities are compared to experiment for the sc (z = 6)
structure, and also to theory for bcc (z = 8) and fcc (z = 12) ferromagnets
in Table 3.3. (Purely magnetic properties which differ for F and AF are not
listed.)

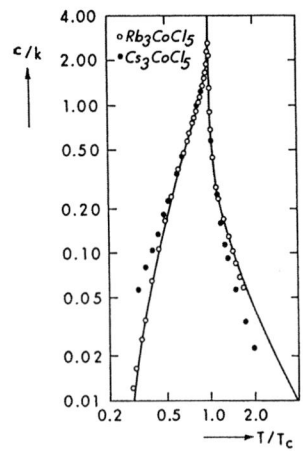

c/k

4.00
2.00
1.00
0.50
0.20
0.10
0.05
0.02
0.01

○ Rb_3CoCl_5
● Cs_3CoCl_5

0.2 0.5 1.0 2.0

⟶ T/T_c

Fig.3.15. *Logarithm* of specific heat for $CoRb_3Cl_5$ and $CoCs_3Cl_5$ ($\cdots$) compared with theory for sc Ising model in 3D [3.38]

Table 3.3. Critical parameters of 3D Ising lattices [3.39]

Lattice	kT_c/zJ	$\mathscr{S}_c/Nk$[a]	$(\mathscr{S}_\infty - \mathscr{S}_c)/Nk$	$-E_0/NkT_c$	$-U_c/NkT_c$
Ising sc z=6	0.7518	0.5579	0.1352	0.6651	0.2200
Ising bcc z=8	0.7942	0.5820	0.1111	0.6296	0.1720
Ising fcc z=12	0.8163	0.5902	0.1029	0.6126	0.1516
$CoRb_3Cl_5$ z=6	0.74	0.563	0.137	0.673	0.226
$CoCs_3Cl_5$ z>6	0.79	0.593	0.106	0.632	0.173

[a]Recall $\mathscr{S}_\infty/Nk = \ln 2 = 0.69315$

The principal sources of reliable information on the 3D Ising model have come from various series expansions. The low T expansion, the high T expansion (there is no known duality linking them in 3D), and critical-point theories combine into a coherent picture which we now attempt to draw, starting with the methodology. The low-T expansions are principally the development of *Sykes, Essam, Gaunt* and various collaborators over the span of a decade [3.40] in a series of papers which the reader should consult for many important features omitted here. High-T series have existed since the dawn of statistical physics, but for present application the papers of *Sykes* et al. [3.41] form a complete documentation.

In the calculation of the free energy and its most sensitive derivatives (c,χ) there is a substantive question of how to extract the discontinuities at T_c and predict the functional form of thermodynamic quantities in a manner most suitable for comparison with experiment and, ultimately, with an exact theory —when such becomes available! Thanks to the efforts of a large number of dilligent researchers, among whom the names of G.A. Baker, Jr., M.F. Sykes, C. Domb, and M.E. Fisher stand out, the analysis of power series —whether in exp(-J/kT) or in tanh(J/kT) —has turned into a refined discipline. An introduction to series analysis has been given by *Gaunt Guttman* [3.42], contributors to the remarkable book, Vol.3 in the series *Phase Transitions and Critical Phenomena* ed. by C. Domb and M.S. Green [3.43]. In addition to the cited paper, this book contains chapters on graph theory (C. Domb), computational techniques for lattice sums (J. Martin), the linked cluster expansion (M. Wortis), the Ising model (C. Domb) and includes the study of other magnetic systems (e.g., Heisenberg and XY models) to which we refer elsewhere in the present text. The information on the 3D Ising model which we now present is, in large measure, called from this sourcebook [3.43] to which the interested reader is referred for full details.

Consider the susceptibility series of the type given in (3.10.25) for the sq lattice. The calculation of the critical exponent 7/4 proceeds easily, because T_c is known exactly. From the equivalent series in 3D, we must have the ability to extract *both* T_c and the relevant exponents, and more. Fortunately, the diagrammatic expansions have been carried out to a large number of terms, with published series such as that for X for the sc lattice

$$X = 1 + 6w + 30w^2 + 150w^3 + 726w^4 + 3510w^5 + 16710w^6 + 79494w^7$$

$$+ 375174w^8 + 1769686w^9 \ldots \tag{3.13.1}$$

and for the face-centered cubic (fcc) lattice:

$$X = 1 + 12w + 132w^2 + 1404w^3 + 14652w^4 + 151116w^5 + 1546332w^6$$

$$+ 15734460w^7 + 159425580w^8 + 1609987708w^9 + \ldots \tag{3.13.2}$$

Coefficients are given in [3.42] to $O(w^{17})$ for the sc, $O(w^{12})$ for the fcc, and $O(w^{15})$ for body-centered cubic (bcc).

By methods that we shall shortly discuss, one extracts from such series the following information:

$$X = A(1 - T_c/T)^{-5/4} \quad (T \to T_c^+) \tag{3.13.3}$$

with A, calculated to 5 decimals, found to be approximately (but not exactly) 1 in all three cubic lattices. T_c, or $w_c = \tanh J/kT_c$, is obtained to

158

better than 1 part in 10^4, as

$$1/w_c(sc) = 4.5844 \text{ , (bcc) } 6.4055 \text{ and (fcc) } 9.8290 \text{ .} \tag{3.13.4}$$

Although the singular behavior (3.13.3) describes the behavior of the power series (3.13.1,2) over a large range of temperatures above T_c, the same thing cannot be said for the specific heat anomaly. After much research, the critical specific heat has been found to be in the form [3.45]

$$c/k = A_\pm(1 - T_c/T)^{-1/8} - B_\pm \text{ .} \tag{3.13.5}$$

At first, series analysis had seemed incapable of providing the critical exponent correctly. (With A_+ and B_+ the parameters above T_c *both* $O(1)$, the constant term remains comparable to the divergent one in the critical region until T is within 1 part in 10^4 of T_c.) Similar considerations apply to A_- and B_- below T_c, and it is an ironic fact that the specific heat anomaly first noted by Kramers, Wannier and Onsager, sparking the intense interest in critical phenomena, remains the hardest to analyze in *all* magnetic models. The susceptibility and other correlation functions are generally much better understood than the specific heat, the critical exponent of which is small (and, in some cases, uncertain even as to sign).

Of course, given a sufficiently long series one can always hope to approach T_c so closely that the leading singularity dominates. In such cases, the critical temperature itself and the critical exponent are calculated as follows. Suppose we know the power series $F(w)$,

$$F(w) = 1 + \sum_{n=1}^{\infty} a_n w^n \tag{3.13.6}$$

for a quantity which should diverge at T_c. Then, presumably,

$$\lim_{n \to \infty} |a_n|^{-1/n} = w_c \tag{3.13.7}$$

fixes w_c. But this rarely occurs; if dominant singularity are $w_c \exp(\pm i\theta)$, a complex pair, the a_n will oscillate asymptotically as

$$a_n \sim \frac{f(n)}{w_c^n} \cos n\theta \tag{3.13.8}$$

with $f(n)$ a slowly varying function. With additional singularities on the complex circle at $|w_c|$ the behavior becomes unpredictable. The *ratio method* is often helpful. It presumes

$$\lim_{n \to \infty} [a_n/f(n)\mu^n] = 1 \tag{3.13.9}$$

with $f(n)$ neither growing nor decaying sufficiently to violate

$$\lim [f(n)]^{1/n} = 1 \quad . \tag{3.13.10}$$

Ultimately, $\mu = 1/w_c$; the question is, how to determine it. It is helpful to assume an asymptotic behavior for $f(n)$:

$$f(n) \sim An^g/g! \tag{3.13.11}$$

which satisfies (3.13.10) and provides us 2 parameters, A and g, with which to help fit the series. We try successive approximations to μ, calculating a_n/a_{n-1}:

$$a_n/a_{n-1} = \mu \left[1 + \frac{g}{n} + O(n^{-2}) \right] \quad . \tag{3.13.12}$$

A plot of a_n/a_{n-1} vs $1/n$ will yield μ as the $1/n = 0$ intercept, with (μg) = slope at that point. Now, a knowledge of g yields the critical index, for the series so defined approaches w_c (i.e., $1/\mu$) as

$$F(w) = A(1 - \mu w)^{-(1+g)} + \text{corrections} \quad , \tag{3.13.13}$$

diverging as specified for $g > -1$ and w positive $\rightarrow w_c$. In Fig.3.16 we display Sykes' simultaneous analyses of the susceptibility and specific heat series for the fcc lattice. The susceptibility series ratios become linear already at small n, whereas the specific heat series ratios become reliable only for $n \gtrsim 10$. The use of *both* series yields a most reliabe $w_c = 1/9.8290$, and critical exponents $\gamma = 5/4$ and $\alpha = 1/8$ [3.44].

If the correction term such as $B_\pm$ in the specific heat is an important quantity, it too can be obtained by analysis of the correction series R(w), once $f(n)$ (A,g,μ) is known:

$$R \equiv \sum_n (a_n - f(n)\mu)w^n \tag{3.13.14}$$

which should be rapidly convergent, or have easily diagnosed singularities.

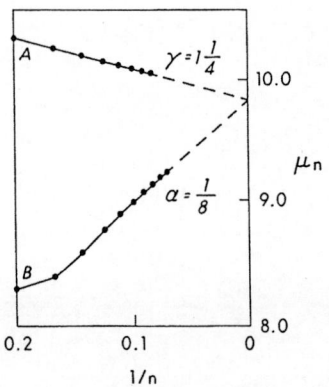

Fig.3.16. Ratios of successive coefficients of the (A) susceptibility and (B) specific heat on fcc lattice, plotted vs 1/n. The extrapolated point yields $1/w_c = 9.8290...$ while the respective slopes yield the critical exponents $\gamma = 5/4$ and $\alpha = 1/8$ [3.44]

When this is not the case, the method of Padé approximants becomes quite useful. Briefly, if a function has singularities such as zeros or poles, it can be written in the form $P(w)/Q(w)$, where P is a polynomial constructed to have its zeros in the desired points, and Q another polynomial having its zeros at the desired poles. We may *approximate* arbitrary functions $F(z)$ by

$$[L,M] = \frac{p_0 + p_1 z + \ldots + p_L z^L}{1 + q_1 z + \ldots + q_M z^M} \tag{3.13.15}$$

called the "L,M Padé approximant", most useful in the study of either $\ln X$, $\ln c/k$ or $X^{1/\gamma}$ or $(c/k)^{1/\alpha}$ in cases γ or α are precisely known. The applications of Padé methods are now quite widespread, and documented in books.

The types of series which have been thus subjected to analysis include high T series such as in the examples above, low T series in $\exp(-J/kT)$, and "density expansions" in which the coefficient of a given power of $\exp(-2B/kT)$ is studied. Both ferromagnets and antiferromagnets have been analyzed, and spins $s > 1/2$ as well. We summarize some of the conclusions:

i) The critical exponents are independent of spin magnitude s. In current jargon, one says that all Ising models of varying s belong to the same "universality class."

ii) The critical exponents *do* depend strongly on d, the dimensionality. Thus, the diamond lattice ($z = 4$) shares critical exponents with the fcc ($z = 12$), rather than with the sq ($z = 4$).

iii) Properties of 3D Ising models are closer to MFT than in 2D, while at or above 4D, the critical behavior closely approximates MFT. As examples, consider the susceptibility exponent $\gamma = 5/4$, approaching the MFT value ($\gamma = 1$), vs. $\gamma = 7/4$ in 2D. The specific heat is anomalous over a smaller interval of temperature in 3D than in 2D; the principal effect of moving through T_c is analogous to the discontinuity found in Chap.2. The critical temperature itself is closer to the MFT prediction. Antiferromagnets in bipartite lattices (sc,bcc) have a parallel susceptibility maximum at $\approx 1.08\ T_c$ rather than at $\approx 1.5\ T_c$ for the sq lattice, closer to the MFT limit which occurs at T_c precisely.

We illustrate with magnetic and thermal data on $DyPO_4$, an analog Ising antiferromagnet on the diamond lattice, $z = 4$ [3.39]. Figure 3.17 illustrates the specific heat-qualitatively that of MFT except near T_c, Fig.3.18 the parallel susceptibility, and Fig.3.19 the sublattice magnetization. It is known that just below T_c the order parameter in 3D Ising models is

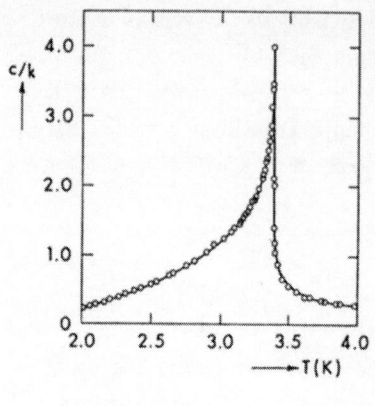

Fig.3.17. Specific heat of Ising spins, $DyPO_4$ on diamond lattice $(z=4)$ data are $(\circ\,\circ\,\circ)$, (——) are from theoretical high- and low-temperature series [3.39]

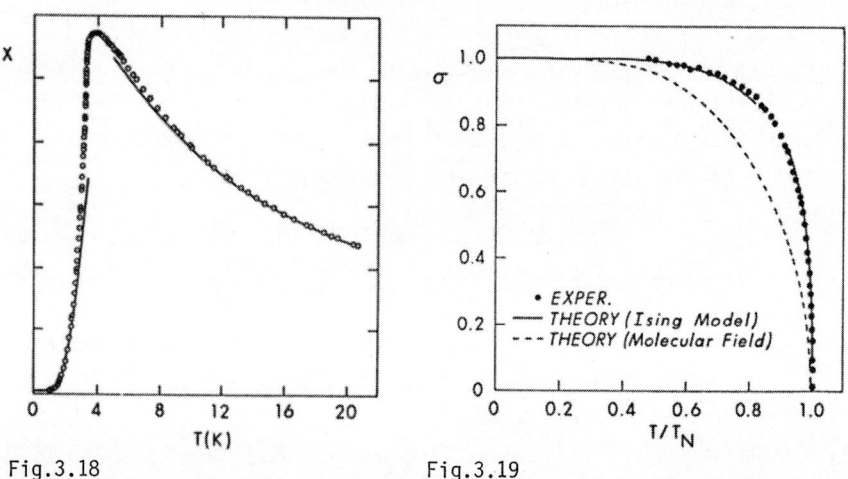

Fig.3.18

Fig.3.19

Fig.3.18. Antiferromagnetic susceptibility $\chi T/\mathbb{C}=X$ vs. T for $DyPO_4$ [3.39]

Fig.3.19. Spontaneous sublattice magnetization (AF order parameter) σ vs. T for $DyPO_4$ [3.39]

$$\sigma(T) = D(1 - T/T_c)^\beta \qquad\qquad (3.13.16)$$

with the magnetization critical exponent (not to be confused with 1/kT!) $\beta = 5/16 = 0.3125$. This should be compared with $\beta = 1/2$ for MFT and $\beta = 1/8$ in the 2D Ising model. The experiment in Fig.3.19 is best fit by $\beta = 0.314$ in excellent agreement with theory.

To conclude, we note that there are more interesting developments in the Ising model than can possibly be recounted here. We have alluded to an extension to higher spin, so let us briefly discuss some properties of higher-spin Ising models before terminating this chapter.

The spin 1 Ising model has $S_i^z = 1,0,-1$, while the spin 3/2 model has $S_i^z = 3/2, 1/2, -1/2, -3/2$, etc. In general, if we wish to consider an Ising model for spins of magnitude $p/2$, the variable $2S_i^z$ can take on the values $p, p-2, \ldots, -p$ and can be written:

$$2S_i^z = \sigma_1 + \sigma_2 + \sigma_3 + \ldots + \sigma_p \qquad (3.13.17)$$

with each individual $\sigma_n = \pm 1$. Thus, the generalized Ising model can be expressed in terms of the Pauli matrices. With this artifice, *Griffiths* [3.46] has extended to arbitrary p the results previously proved for $p = 1$, such as the Lee-Yang theorem [3.6] that the zeros of the partition function lie on the unit circle in the complex $\exp(-2B/kT)$ plane. He also proved an inequality for arbitrary p (Griffiths' inequality) [3.47]: If A and B are spins or products thereof,

$$<AB> \geq <A><B> \qquad (3.13.18)$$

in ferromagnets. For $p \neq 1$, one can also include crystal field effects via $(S_i^z)^2$ type terms, and consideration of these has led to the identification of several new phase transitions, some of first-order. We cite a few representative titles in [3.48].

With these remarks, we end this text, hoping it has introduced the vast field of magnetism — the known literature and the vast unknown of future research — to the reader. Let the oxen beware! [3.49].

References

Chapter 1

1.1 A.B. Pippard: *Elements of Classical Thermodynamics* (Cambridge Univ. Press, London 1966) p.65
1.2 D.C. Mattis: *The Theory of Magnetism I, Statics and Dynamics*, Springer Ser. Solid State Sci., Vol.17 (Springer, Berlin, Heidelberg 1981)
1.3 D.J. Amit: *Field Theory, Renormalization Group and Critical Phenomena* (McGraw-Hill, New York 1978)
   J.B. Kogut: An introduction to lattice gauge theory and spin systems, Rev. Mod. Phys. **51**, 659 (1979)
   R. Savit: Duality in field theory and statistical systems, Rev. Mod. Phys. **52**, 453 (1980)
1.4 H.E. Stanley: *Introduction to Phase Transitions and Critical Phenomena* (Oxford Univ. Press, London 1971)
   A more recent treatment was given by
   P. Pfeuty, G. Toulouse: *Introduction to the Renormalization Group and to Critical Phenomena* (Wiley, London, New York 1977)
1.5 S.-K. Ma: *Modern Theory of Critical Phenomena* (Benjamin, Reading, MA 1976)
1.6 K. Binder (ed.): *Monte Carlo Methods in Statistical Physics*, 2nd ed. Topics Current Phys., Vol.7 (Springer, Berlin, Heidelberg 1985)
   K. Binder (ed.): *Applications of the Monte Carlo Method in Statistical Physics*, Topics Current Phys., Vol.36 (Springer, Berlin, Heidelberg 1984)

Chapter 2

2.1 D.C. Mattis: *The Theory of Magnetism I*, Springer Ser. Solid State Sci., Vol.17 (Springer, Berlin, Heidelberg 1981)
2.2 M.P. Langevin: J. Physique **4**, 678 (1905)
2.3 P. Weiss: *Proc. of 6th Solvay Congress, 1930* (Gauthier-Villars, Paris 1932)
2.4 L. Brillouin: J. Phys. Radium **8**, 74 (1927)
2.5 P. Weiss divined the existence of domains: J. Physique [4] **6**, 661 (1907)
2.6 Domains are thoroughly described in a number of texts:
   R.M. Bozorth: *Ferromagnetism* (Van Nostrand, Princeton 1951)
   A.H. Morrish: *The Physical Principles of Magnetism* (Wiley, New York 1964)
   S. Chikazumi: *Physics of Magnetism* (Wiley, New York 1964)
   A.H. Eschenfelder: *Magnetic Bubble Technology*, 2nd ed., Springer Ser. Solid State Sci., Vol.14 (Springer, Berlin, Heidelberg 1981)
2.7 L. Néel: Ann. Physique **17**, 64 (1932)

2.8 F. Bitter: Phys. Rev. **54**, 79 (1937)
2.9 J.H. VanVleck: J. Chem. Phys. **9**, 85 (1941)
2.10 J.H. VanVleck: Rev. Mod. Phys. **17**, 27 (1945), p. 45ff.
2.11 M.H. Cohen, F. Keffer: Phys. Rev. **99**, 1128 and 1135 (1955)
     R.B. Griffiths: Phys. Rev. **176**, 655 (1968)
     M.E. Fisher, D. Ruelle: J. Math. Phys. **7**, 260 (1966)
2.12 M.A. Ruderman, C. Kittel: Phys. Rev. **96**, 99 (1954). For details see
     [Ref.2.1, Chaps.2 and 6]
2.13 D.C. Mattis, W. Donath: Unpublished IBM report (ca. 1961). For
     summary see [Ref.2.1, p.235 and Fig.6.6]
2.14 G.H. Wannier: Phys. Rev. **79**, 357 (1950)
2.15 G. Toulouse: Commun. Phys. **2**, 115 (1977)
     E. Fradkin, B. Huberman, S. Shenker: Phys. Rev. B**18**, 4789 (1978)
2.16 H.R. Ott et al.: Phys. Rev. B**25**, 477 (1982)
2.17 T. Berlin, M. Kac: Phys. Rev. **86**, 821 (1952)
2.18 G.S. Joyce: J. Phys. A**5**, L65 (1972);
     M.L. Glasser: J. Math. Phys. **13**, 1145 (1972);
     M.L. Glasser, I.J. Zucker: Proc. Natl. Acad. Sci. USA **74**, 1800
     (1977) [NB: following their Eq. (8b), $I_3(3)$ should be divided by
     384π.]
2.19 S. Katsura, T. Morita, S. Inawashiro, T. Horiguchi, Y. Abe: J. Math.
     Phys. **12**, 892 (1971);
     S. Katsura, S. Inawashiro, Y. Abe: J. Math. Phys. **12**, 895 (1971)
2.20 J. Kosterlitz, D. Thouless, R.C. Jones: Phys. Rev. Lett. **36**, 1217
     (1976)
2.21 D. Mattis, R. Raghavan: Phys. Lett. **75A**, 313 (1980)
2.22 M.L. Mehta: *Random Matrices* (Academic, New York 1967) Appendix A29
2.23 The nearest-neighbor random bond model is treated in the spherical
     model approximation by L. deMenezes, A. Rauh, S.R. Salinas: Phys.
     Rev. B**15**, 3485 (1977). A quantum version of the spherical spin glass
     is in P. Shukla and S. Singh: Phys. Lett. **81A**, 477 (1981). A peculiar
     failure of the spherical model (the spurious phase transition of a
     single spin!) has been discovered by E.H. Lieb, C.J. Thompson: J.
     Math. Phys. **10**, 1403 (1969). The AF spherical model was first examined
     by R. Mazo: J. Chem. Phys. **39**, 2196 (1963)
2.24 D. Sherrington, S. Kirkpatrick: Phys. Rev. Lett. **35**, 1792 (1975), and
     Phys. Rev. B**17**, 4384 (1978). Despite many claims to the contrary, this
     most elementary molecular-field type spin glass has not been solved to
     date, except in the spherical-model version [2.20]; see [2.32]
2.25 S.F. Edwards, P.W. Anderson: J. Phys. F**5**, 965 (1975)
2.26 L.R. Walker, R.E. Walstedt: Phys. Rev. B**22**, 3816 (1980)
2.27 N.D. Mackenzie, A.P. Young: Phys. Rev. Lett. **49**, 301 (1982)
2.28 G. Parisi: J. Phys. A**13**, 1101, 1887, L115 (1980) and Phil. Mag. B**41**,
     677 (1980)
2.29 H. Sompolinsky: Phys. Rev. Lett. **47**, 935 (1981)
2.30 T. Jonsson: Phys. Lett. **91A**, 185 (1982)
     C. deDominicis, M. Gabay, C. Orland: J. Physique Lett. **42**, L523 (1981)
2.31 A.J. Bray, M.A. Moore: J. Phys. C**12**, L441 (1979)
     F. Bantilan, R.G. Palmer: J. Phys. F**11**, 261 (1981)
2.32 Some 700 recent titles on the topic of mean-field spin glasses are re-
     viewed by D. Chowdhury, A. Mookerjee: Phys. Rpts. **114**, 1-98 (1984)
2.33 F. Keffer: Spin Waves, in *Handbuch d. Physik*, XVIII/2, ed. by H.J.P.
     Wijn (Springer, Berlin 1966) pp.1-273; cf. his Eq. (9.21)ff.
2.34 F. Holtzberg et al.: J. Appl. Phys. **35/2**, 1033 (1964)
2.35 M. Bloch: Phys. Rev. Lett. **9**, 286 (1962); J. Appl. Phys. **34**, 1151
     (1963). This work is extended and illuminated by I. Goldhirsch and
     V. Yakhot: Phys. Rev. B**21**, 2833 (1980)
2.36 N.D. Mermin, H. Wagner: Phys. Rev. Lett. **17**, 1133, 1307 (1966)

2.37 H.E. Stanley, T. Kaplan: Phys. Rev. Lett. **17**, 913 (1966)
The proof of the existence of such a phase transition, for the plane
rotator (classical XY) model is in J. Fröhlich, T. Spencer: Phys. Rev.
Lett. **46**, 1006 (1981), based on the general method of J. Fröhlich,
E.H. Lieb: Commun. Math. Phys. **60**, 233 (1978)
2.38 L.J. deJongh, A.R. Miedema: Adv. Phys. **23**, 1 (1974) treated Heisenberg
antiferromagnets in 2D, p.64ff.
2.39 A Monte-Carlo renormalization-group analysis by S. Shenker, J. Tobochnik: Phys. Rev. B**22**, 4462 (1980)
2.40 N.N. Bogoliubov: Phys. Abh. Sowj. **6**, 1, 113, 229 (1962)
H. Wagner: Z. Phys. **195**, 273 (1966)
2.41 M.E. Fisher, D. Jasnow: Phys. Rev. B**3**, 907 (1971)
2.42 J. Fröhlich, E.H. Lieb: Phys. Rev. Lett. **38**, 440 (1977); Commun.
Math. Phys. **60**, 233 (1978)
2.43 F.J. Dyson, E.H. Lieb, B. Simon: J. Stat. Phys. **18**, 335 (1978)
2.44 B. Simon: Phys. Rev. Lett. **44**, 547 (1980)
2.45 L. Fadeev, L. Takhtajan: Phys. Lett. **85A**, 375 (1981)
2.46 K. Nakamura, T. Sasada: J. Phys. C**15**, L1013 (1982)
2.47 A. Caliri, D.C. Mattis: Rev. Brasileira de Fisica **13**, 322 (1983)
V. Glauss, T. Schneider, E. Stoll: Phys. Rev. B**27**, 6770 (1983)
2.48 J. Groen, T. Klaasen, N. Poulis, G. Müller, H. Thomas, H. Beck:
Phys. Rev. B**22**, 5369 (1980). This work was extended by J. Kurmann,
H. Thomas, G. Müller: Physica **112A**, 235 (1982)
2.49 M.E. Fisher: Am. J. Phys. **32**, 343 (1964)
2.50 J. Oitmaa, D.D. Betts, L.G. Marland: Phys. Lett. **79A**, 193 (1980)
(Ground state)
D.D. Betts, F. Salevsky, J. Rogiers: J. Phys. A**14**, 531 (1981) (Vortex
operators)
J. Rogiers, T. Lookman, D.D. Betts, C.J. Elliott: Can. J. Phys. **56**,
409 (1978) (High T series expansions)
2.51 D.C. Mattis: Phys. Rev. Lett. **42**, 1503 (1979) (Proof that ground
state in XY model has $S_{tot}^Z = 0$.) Related work on other models was
carried out by D. Mattis: Phys. Rev. **130**, 76 (1963);
E. Lieb, D. Mattis: J. Math. Phys. **3**, 749 (1962)
2.52 A. Lagendijk, H. De Raedt: Phys. Rev. Lett. **49**, 602 (1982)
P.M. Grant, E. Loh, Jr., D. Scalapino: Phys. Rev. **B** (to appear in "Rapid Commun.", 1985)
2.53 J.M. Kosterlitz, D.J. Thouless: J. Phys. C**6**, 1181 (1973); and similarly, V.L. Berezisnkii: Sov. Phys. JETP **32**, 493 (1971)
2.54 J.M. Kosterlitz: J. Phys. C**7**, 1046 (1974)
2.55 J. Villain: J. Physique **36**, 581 (1975)
2.56 J. José, L. Kadanoff, S. Kirkpatrick, D. Nelson: Phys. Rev. B**16**, 1217
(1977)
2.57 J. Tobochnick, G.V. Chester: Phys. Rev. B**20**, 3761 (1979)
2.58 The first attempt at a transfer matrix in this problem seems to be that
of A. Luther, D. Scalapino: Phys. Rev. B**16**, 1153 (1977), who truncate
it into the form of a s =1 chain in an external field: $\Delta(S_i^z)^2$. Such
chains have been most recently studied by S.T. Chui, K.B. Ma: Phys.
Rev. B**29**, 1287 (1984), and it seems almost certain they *can* yield the
general features of the plane rotator model. However, the derivation
of the transfer matrix in the present book seems to be the first rigorous one.
2.59 The procedure: mapping of the transfer matrix onto the s =1/2 anisotropic Heisenberg antiferromagnetic linear chain, has been given in D.
Mattis: Phys. Lett. **104**, 357 (1984)
2.60 M. Abramowitz, I. Stegun (eds.): *Handbook of Math. Functions* (National
Bureau of Standards, Washington 1964). Sections 9.6,7

Chapter 3

3.1 S.G. Brush: The history of the Lenz-Ising model: Rev. Mod. Phys. **39**, 883 (1967)

3.2 W. Lenz: Phys. Z. **21**, 613 (1920)

3.3 E. Ising: Z. Physik **31**, 253 (1925)

3.4 H. Bethe: Proc. Roy. Soc. (London) A**150**, 552 (1935); J. Appl. Phys. **9**, 244 (1938). Also: F. Cernuschi, H. Eyring: J. Chem. Phys. **7**, 547 (1939)

3.5 R. Fowler, E. Guggenheim: *Statistical Thermodynamics* (Cambridge Univ. Press, Cambridge 1939) Chap.13

3.6 C.N. Yang, T.D. Lee: Phys. Rev. **87**, 404 (1952)
T.D. Lee, C.N. Yang: Phys. Rev. **87**, 410 (1952)
Recent work on the complex zeros includes:
E. Marinari: Nucl. Phys. B**235** (FS11), 123 (1984), 3D Ising model
K. De'Bell, M.L. Glasser: Phys. Lett. **104**A, 255 (1984), Cayley tree
W. Saarloos, D. Kurtze: J. Phys. A**17**, 1301 (1984), Ising model
A. Caliri, D. Mattis: Phys. Lett. **106**A, 74 (1984), long-range model of (2.7.2) with $J_0 \gtreqless 0$

3.7 R. Peierls: Proc. Camb. Phil. Soc. **32**, 477 (1936)

3.8 R. Griffiths: Phys. Rev. **136**, A437 (1964). The reader will find it instructive to determine where this proof fails for the XY model!
See further corrections and extension in
C.-Y. Weng, R. Griffiths, M. Fisher: Phys. Rev. **162**, 475 (1967)

3.9 L. Onsager: Phys. Rev. **65**, 117 (1944), algebraic formulation
B. Kaufman: Phys. Rev. **76**, 1232 (1949), spinor reformulation
L. Onsager: Nuovo Cimento (Suppl.) **6**, 261 (1949), spontaneous magnetization;
C.N. Yang: Phys. Rev. **85**, 809 (1952), first derivation of Onsager's formula for magnetization in the literature

3.10 H. Kramers, G. Wannier: Phys. Rev. **60**, 252, 263 (1941)

3.11 D.C. Mattis: *The Theory of Magnetism* I, Springer Ser. Solid-State Sci., Vol.17 (Springer, Berlin Heidelberg 1981)

3.12 C. Domb: On the Theory of Cooperative Phenomena in Crystals, Adv. Phys. **9**, 149-361 (1960). The fit of $T_c(d)$ on hypercubic lattices to 2 straight lines was performed by
G. Cocho, G. Martinez-Mekler, R. Martinez-Enriquez: Phys. Rev. B**26**, 2666 (1982)

3.13 M.E. Fisher: Phys. Rev. **162**, 480 (1967)

3.14 H.R. Ott et al.: Phys. Rev. B**25**, 477 (1982);
Z. Chen, M. Kardar: Phys. Rev. B**30**, 4113 (1984)

3.15 M.E. Lines: Phys. Rpts. **55**, 133 (1979)

3.16 E. Jahnke, F. Emde: *Tables of Functions* (Dover, New York 1945)

3.17 T.A. Tjon: Phys. Rev. B**2**, 2411 (1970)
B. McCoy, J. Perk, R. Schrock: Nucl. Phys. B**220**, 35, 269 (1983) and references therein

3.18 E. Lieb, T. Schultz, D. Mattis: Ann. Phys. (NY) **16**, 407 (1961)

3.19 P. Pfeuty: Phys. Lett. **72**A, 245 (1979)

3.20 T. Schultz, D. Mattis, E. Lieb: Rev. Mod. Phys. **36**. 856 (1964)

3.21 A review of Toeplitz matrices, and various improvements and applications thereof to statistical mechanics has been published by M. Fisher, R. Hartwig: Adv. Chem. Phys. **15**, 333-354 (1968). The original application to the Ising model in the familiar literature seems to be E. Montroll, R. Potts, J. Ward: J. Math. Phys. **4**, 308 (1963) in the Onsager anniversary issue of that Journal. But Montroll et al. disclaim first use, and credit Onsager:

... this is one of the methods used by Onsager himself. Mark Kac alerted the authors to a limit formula for the calculation of large Toeplitz determinants which appear naturally in the theory of spin correlations in a two-dimensional Ising lattice. This formula was first discussed by Szegö [Comm. Séminaire Math. Univ. Lund, tome suppl. (1952) dédié à M. Riesz, p. 228]. Perusal of the Szegö paper shows that the problem was proposed to Szegö by the Yale mathematician S. Kakutani, who apparently heard it from Onsager...

3.22 T. Oguchi: J. Phys. Soc. Jpn. **6**, 31 (1951)
3.23 M.F. Sykes: J. Math. Phys. **2**, 52 (1961)
3.24 G.A. Baker, Jr.: Phys. Rev. **124**, 768 (1961)
3.25 E. Barouch, B. McCoy, T.T. Wu: Phys. Rev. Lett. **31**, 1409 (1973)
3.26 M. Plischke, D. Mattis: Phys. Rev. B**2**, 2660 (1970)
3.27 E. Barouch: Physica 1D, 333 (1980)
     Generalizations of the Lee-Yang methods [3.6] to other models have recently appeared, notably:
     M. Bander, C. Itzykson: Phys. Rev. B**30**, 6485 (1984) for O(N) spin models
     D. Kurtze, M. Fisher: J. Stat. Phys. **19**, 205 (1978) for spherical models
3.28 R. Baxter, I. Enting: J. Phys. A**11**, 2463 (1978)
3.29 G. Wannier: Phys. Rev. **79**, 357 (1950)
3.30 T. Utiyama: Progr. Theor. Phys. **6**, 907 (1951)
3.31 M. Sykes, M. Fisher: Physica **28**, 919, 939 (1962)
3.32 E. Lieb, D. Ruelle: J. Math. Phys. **13**, 781 (1972)
3.33 E. Müller-Hartmann, J. Zittartz: Z. Physik B**27**, 261 (1977)
3.34 J. Zittartz: Z. Physik B**40**, 233 (1980)
3.35 K.Y. Lin, F.Y. Wu: Z. Physik B**33**, 181 (1979)
3.36 A. Bienenstock: J. Appl. Phys. **37**, 1459 (1966)
3.37 M. Plischke, D.C. Mattis: Phys. Rev. A**3**, 2092 (1971)
3.38 H. Blöte, W. Huiskamp: Phys. Lett. A**29**, 304 (1969)
3.39 L. de Jongh, A. Miedema: Experiments on Simple Magnetic Model Systems, Adv. Phys. **23**, 1-260 (1974)
3.40 M. Sykes, J. Essam, D. Gaunt: J. Math. Phys. **6**, 283 (1965)
     M. Sykes, D. Gaunt, J. Essam, D. Hunter: J. Math. Phys. **14**. 1060 (1973)
     M. Sykes, D. Gaunt, S. Mattingly, J. Essam, C. Elliott: J. Math. Phys. **14**, 1066 (1973)
     M. Sykes, D. Gaunt, J. Martin, S. Mattingly, J. Essam: J. Math. Phys. **14**, 1071 (1973)
     M. Sykes, D. Gaunt, J. Essam, B. Heap, C. Elliott, S. Mattingly: J. Phys. A**6**, 1498 (1973)
     M. Sykes, D. Gaunt, J. Essam, C. Elliott: J. Phys. A**6**, 1506 (1973)
     D. Gaunt, M. Sykes: J. Phys. A**6**, 1517 (1973)
3.41 M. Sykes, D. Gaunt, P. Roberts, J. Wyles: J. Phys. A**5**, 624, 640 (1972)
     M. Sykes, D. Hunter, D. McKenzie, B. Heap: J. Phys. A**5**, 667 (1972)
3.42 D. Gaunt, J. Guttmann: Asymptotic Analysis of Coefficients, in *Phase Transitions and Critical Phenomena*, Vol.3, ed. by C. Domb and M. Green (Academic, New York 1974)
3.43 C. Domb, M. Green (eds.): *Phase Transitions and Critical Phenomena*, Vol.3 (Academic, New York 1974)
3.44 M. Sykes et al.: J. Phys. A**5**, 640 (1972) Appendix
3.45 See the recent analysis and references in
     S. Jensen, O. Mouritsen: J. Phys. A**15**, 2631 (1982) or [3.43]
3.46 R.B. Griffiths: J. Math. Phys. **10**, 1559 (1969)
3.47 R.B. Griffiths: J. Math. Phys. **8**, 478, 484 (1967)
3.48 M. Blume: Phys. Rev. **141**, 517 (1966)
     H.W. Capel: Physica **37**, 423 (1967) and references therein (Blume-Capel model)
     H. Chen, P.M. Levy: Phys. Rev. B**7**, 4267 (1973)
     D. Furman, S. Dattagupta, R.B. Griffiths: Phys. Rev. B**15**, 441 (1977)
     E.K. Riedel, F.J. Wegner: Phys. Rev. B**9**, 294 (1974)
     G.B. Taggart: Phys. Rev. B**20**, 3886 (1979)
3.49 When Pythagoras established the theorem of the square upon the hypothenuse he sacrificed 1000 oxen to Apollo. Since then, whenever anyone has had a new idea, oxen everywhere have trembled

List of Tables

Subject Index

AF (antiferromagnet; antiferromagnetism) 26,147

bcc (body-centered cubic lattice) 100,157

Bessel functions

$I_n(K)$ 66,76,85-87

$j_\ell(i\beta J)$ 78

β (= 1/kT) 10

Bogolubov transformation 117,138

Bosons 35,64

Brillouin function 22

c (specific heat, heat capacity) 1,15

as a fluctuation in energy 14

discontinuity 20,135,136

in anisotropic Heisenberg chain 75

in 1D Ising model 106

in 2D plane rotator 84

in anisotropic 2D Ising model 135

in 3D Ising model 157,160-162

in MFT 20ff.

in spherical model 45,56

χ (or $\chi_0 = \chi(B = 0)$), as a fluctuation in $\mathcal{M}$ 15,93,102,107,137

in antiferromagnets 28,47,51,52, 77,78,79,105,106,115,142,153, 154,162

in ferromagnets 24,26,47,49,50, 61,77,78,93,105,106,137ff.,142, 153

Configuration average 53,55

Critical exponents 44,49,50,83,141-143,153,154,158-162

Crystal field effects 163

Cumulants 92

Curie constant 16,26

Cyclic determinant 139

Decay rate 118

Density matrix 13,129

Density of states 54,56,59

Dual lattice 82,98,99

Duality 97ff.,110,127,147

Elliptic integrals 116

Extensive variables 9

F (free energy), f (free energy/spin) 12,15,39,40,42,53,55,93,133-135, 144,152

fcc (face-centered-cubic lattice) 100,157

Fermions 34,72,73ff.,111-114,120ff., 130

Fluctuations 12

Frustration 32,152

Gaussian integrals 36

Gaussian model 38-41,47,48,50,134, 135

Graphs 95ff.,158

Griffiths inequality 163

173

U (internal energy), u (internal energy/spin) 15,19,45,74,135,136, 137,157

Vortex 80ff.

W(d,τ) (generalized Watson's integral) 40,46

Weiss field *see* Molecular field

Wigner's distribution (eigenvalues of random matrix) 55,56,59

XY model 73

Zassenhaus formula 131

Errata for The Theory of Magnetism I
(Springer Series in Solid-State Sciences, Vol. 17)

Ignoring a number of obvious misprints, the following changes should be brought to this book:

At end of Sect. 1.8, p.30, insert: "The solitons discussed in Chap.5 are examples of domains. A practical application of domains to magnetic bubbles is given in the following section."

On the first line of p.84, replace "+" by "-".

On the second line above (3.87), p.89, change "$\leq$" to "$\geq$".

In Problem 4.5, p.109, replace subscript "11" by "1".

Equation (5.109), p.162, should read:

$$a_i \rightarrow (1 - n_i/2s)^{-\frac{1}{2}} a_i \quad \text{and} \quad a_i^* \rightarrow a_i^* (1 - n_i/2s)^{\frac{1}{2}}$$

Line below (5.146), p.175, should read: "Other limits which can be studied by related methods include:"

Figure 5.8, p.179: $q = ka$, where a is the nearest-neighbor distance.

Page 206, next-to-last paragraph in the Sect. 5.17, starting "Although there is.... replace by: "In 3D, the solitons are domain walls. Their energy $O(N^{2/3})$ is compensated by the magnetostatic energy reduction. Thus, solitons *alias* domains are required by thermodynamics at finite temperature."

In (6.112), p.255, and (6.127), p.263, change (9/5) to (9/10).

D.C.Mattis

The Theory of Magnetism I

Statics and Dynamics

1981. 58 figures. XV, 300 pages. (Springer Series in
Solid-State Sciences, Volume 17)
ISBN 3-540-10611-1

Starting with a thorough historical introduction to
the study of magnetism – one of the oldest
sciences known to man – through the most
modern developments (magnetic "bubbles" and
"soap films", effects of magnetic impurities in
metals and "spin glasses"), this book develops the
concepts and the mathematical expertise necessary
to understand contemporary research in this field.

Volume I treats exchange forces, the theory of
angular momentum, proves important theorems
concerning the nature of the ground state and
excited states, develops theories of magnons,
vortices and solitons and gives a survey of the
rapidly evolving field of magnetism in metals. The
approach is thorough: all important theories are
worked out in detail, using methods and notation
that are uniform throughout. Footnotes and biblio-
graphy provide a guide to the original literature,
and a number of problems test the reader's skill.

Springer-Verlag
Berlin
Heidelberg
New York
Tokyo

R. M. White

Quantum Theory of Magnetism

2nd corrected and updated edition. 1983.
113 figures. XI, 282 pages. (Springer Series in
Solid-State Sciences, Volume 32)
ISBN 3-540-11462-9
(Originally published by McGraw-Hill, Inc.,
New York, 1970)

Contents: The Magnetic Susceptibility. – The
Magnetic Hamiltonian. – The Static Susceptibility
of Noninteracting Systems. – The Static Suscepti-
bility of Interacting Systems. – The Dynamic
Susceptibility of Weakly Interacting Systems. –
The Dynamic Susceptibility of Strongly Interacting
Systems. – Magnetic Impurities. – Neutron Scatter-
ing. – References. – Subject Index.

This monograph presents a uniform treatment of
the physical principles underlying magnetic phenomena in matter in the context of linear response
theory. This provides the reader with a conceptual
framework for understanding the wide range of
magnetic phenomena as well as the relationship of
magnetism to other areas of physics. Since the first
edition was published in 1970 numerous advances
have occured and new phenomena have been
discovered.

These new developments have been incorporated
into the original organization thereby bringing the
book up to date as well as indicating the generality
of the approach.

Springer-Verlag
Berlin
Heidelberg
New York
Tokyo

Springer Series in Solid-State Sciences

Editors: M. Cardona P. Fulde H.-J. Queisser